Energy Sustainability and Efficiency: Global Perspectives and Innovative Approaches

RIVER PUBLISHERS SERIES IN ENERGY SUSTAINABILITY AND EFFICIENCY

Series Editors:

PEDRAM ASEF
Lecturer (Asst. Prof.) in Automotive Engineering
University of Hertfordshire, UK

The "River Publishers Series in Sustainability and Efficiency" is a series of comprehensive academic and professional books which focus on theory and applications in sustainable and efficient energy solutions. The books serve as a multi-disciplinary resource linking sustainable energy and society, fulfilling the rapidly growing worldwide interest in energy solutions. All fields of possible sustainable energy solutions and applications are addressed, not only from a technical point of view, but also from economic, social, political, and financial aspects. Books published in the series include research monographs, edited volumes, handbooks and textbooks. They provide professionals, researchers, educators, and advanced students in the field with an invaluable insight into the latest research and developments.

Topics covered in the series include, but are not limited to:

- Sustainable energy development and management;
- Alternate and renewable energies;
- Energy conservation;
- Energy efficiency;
- Carbon reduction;
- Environment.

For a list of other books in this series, visit www.riverpublishers.com

Energy Sustainability and Efficiency: Global Perspectives and Innovative Approaches

Editors

Momir Tabakovic

University of Applied Sciences Technikum Wien, Austria

Svitlana Alyokhina

University of Applied Sciences Technikum Wien, Austria

River Publishers

Routledge
Taylor & Francis Group
NEW YORK AND LONDON

Published 2026 by River Publishers
River Publishers
Broagervej 10, 9260 Gistrup, Denmark
www.riverpublishers.com

Distributed exclusively by Routledge
605 Third Avenue, New York, NY 10017, USA
4 Park Square, Milton Park, Abingdon, Oxon OX14 4RN

Energy Sustainability and Efficiency: Global Perspectives and Innovative Approaches / by Momir Tabakovic, Svitlana Alyokhina.

Routledge is an imprint of the Taylor & Francis Group, an informa business

ISBN 978-87-4380-684-4 (hardback)
ISBN 978-87-4380-687-5 (paperback)
ISBN 978-87-4380-686-8 (online)
ISBN 978-87-4380-685-1 (master ebook)
DOI: 10.1201/9788743806851

Contents

Preface

M. Tabakovic and S. Alyokhina

University of Applied Sciences Technikum Wien, Department Industrial Engineering, Austria
E-mail: momir.tabakovic@technikum-wien.at; svitlana.alyokhina@technikum-wien.at

In recent years, the global energy sector has undergone remarkable changes: the total global renewable energy capacity has reached 3,869 GW,[1] which is nearly 14% higher than the previous year (with the growth rate increasing each year). It is interesting to note that in Europe, more than 45% of electricity is already generated from clean sources, whereas in developing countries, this figure remains below 30%,[2] highlighting the importance of a global and balanced approach to sustainable energy development.

Energy-sustainable development is a concept aimed at ensuring long-term energy security, economic stability, and environmental sustainability. The modern world faces numerous challenges, including dependence on fossil fuels, increasing energy consumption, uneven distribution of energy resources, and the negative impact of traditional energy sources on the environment. In this context, energy-sustainable development encompasses a wide range of strategies and technologies designed to create a balanced, accessible, and environmentally safe energy system.

Amidst the global climate crisis and the depletion of traditional energy resources, the search for effective solutions for sustainable energy supply is becoming increasingly relevant. Achieving energy sustainability is possible only through a comprehensive approach that includes the development of renewable energy sources, improving energy efficiency at both the equipment level and within energy systems as a whole, digitizing energy infrastructure, and actively involving society in the energy transition process. A key element

[1]IRENA (2024), Renewable capacity statistics 2024, International Renewable Energy Agency, Abu Dhabi.

[2]Renewable Energy Progress Tracker. IEA. https://www.iea.org/data-and-statistics/data-tools/renewable-energy-progress-tracker (Accessed 7 February 2025)

of this process is the decentralization of the energy system, which reduces dependence on centralized power plants and ensures a more even distribution of energy (especially for residential consumers).

Nowadays, one of the most significant areas of energy consumption is the building sector, which accounts for nearly 40% of global energy use and contributes substantially to greenhouse gas emissions.[3] Improving energy efficiency in buildings is therefore a crucial step toward achieving sustainable energy goals. Meanwhile achieving meaningful progress in building energy efficiency requires a holistic approach that considers both technical and social factors. The successful implementation of energy-efficient strategies depends not only on technological advancements but also on user behavior, community engagement, and policy support.

The book "*Energy Sustainability and Efficiency: Global Perspectives and Innovative Approaches*" highlights some of the innovative approaches to enhancing energy sustainability and efficiency on a global scale with a particular focus on the building sector. It brings together scientific research, technological advancements, and socioeconomic strategies to present best practices and promising solutions for a fair energy transition, digitization, and automation in the building energy sector.

This book consists of six chapters, each focusing on a distinct innovative aspect of the global efforts toward sustainable development and energy efficiency. At the same time, the editors acknowledge that it is impossible to comprehensively cover all existing areas of energy efficiency and sustainable development within a single book. The emergence of new, cutting-edge innovations in renewable energy and related fields aimed at achieving decarbonization and sustainability goals will likely serve as an impetus for compiling a future volume.

The *first chapter* titled "Community Engagement for PED Development and Just Energy Transition: A Research Protocol" explores the concept of positive energy districts (PEDs) and their role in urban energy transitions. The chapter begins by highlighting the critical responsibility of cities in addressing climate change due to their high energy consumption and carbon emissions. It introduces PEDs as urban areas that generate more renewable energy than they consume, thereby aiming for zero CO_2 emissions and contributing to the European Union's climate neutrality goals. However, the

[3] Buildings – Energy System – IEA. https://www.iea.org/energy-system/buildings (Accessed 24 March 2025)

implementation of PEDs faces significant challenges, particularly in ensuring equitable energy transitions.

The chapter focuses on the results of the Citizens4PED[4] project, which integrates socio-cultural, institutional, and technical dimensions in PED development. The core approach involves the establishment of Living Labs – participatory environments where local stakeholders, including vulnerable groups, collaborate on energy solutions. The project's research protocol aims to enhance stakeholder engagement and community empowerment, ensuring a just transition in diverse urban contexts.

A key methodological component discussed is stakeholder analysis, a well-established approach in participatory planning. The chapter details a five-step research protocol designed for stakeholder engagement in PED development. The chapter also explores social network analysis as a tool for mapping stakeholder relationships, identifying key actors, and fostering collaboration.

In conclusion, the chapter argues that community-led energy transitions are essential for achieving sustainable and just urban development. By integrating stakeholder engagement methodologies with participatory action research, the research protocol contributes to the broader PED discourse, providing practical strategies for fostering local energy transition partnerships. While the Living Labs are still in their early stages, initial findings suggest that the protocol effectively fosters dialogue, reveals stakeholder interests, and strengthens community participation. Ultimately, the chapter advocates for an approach that prioritizes social equity alongside technical and regulatory advancements, ensuring that PEDs serve as inclusive solutions for the climate crisis.

As mentioned earlier, the transition to a sustainable energy future is one of the most pressing global challenges of our time. As societies worldwide grapple with the urgent need to reduce carbon emissions and shift toward renewable energy sources, innovative solutions have emerged to facilitate this transformation. Among these, Energy Communities[5] have gained increasing prominence as a model that empowers individuals and local groups to participate actively in the energy transition. The *second chapter* "Revolution or Illusion? Austrian Visions and Realities of Energy Communities" delves into the rise of energy communities in Austria, assessing their effectiveness,

[4]https://citizens4ped.eu/

[5]https://energy.ec.europa.eu/topics/markets-and-consumers/energy-consumers-and-prosumers/energy-communities_en#related-links

the motivations of their members, and the barriers that hinder their broader adoption.

The chapter provides a detailed analysis of the motivations driving individuals to join energy communities. Many citizens perceive them as an opportunity to lower their electricity costs through reduced grid fees and shared renewable energy production. Additionally, environmental motivations play a crucial role, as Energy Community members seek to contribute to a more sustainable and carbon-neutral energy system. The ability to achieve a degree of energy autonomy – reducing dependence on large utility providers – is another significant incentive. Despite these benefits, several barriers impede the widespread adoption of energy communities: administrative complexity; the bureaucratic establishment process, and the difficulty and security of navigating these processes.

The study presented in this chapter employs an importance-performance analysis based on survey data from 1,073 Austrian residents. This methodological approach allows the authors to assess how well energy communities align with public expectations by comparing perceived importance with actual performance across various dimensions. Furthermore, the chapter calls for a reassessment of the regulatory framework governing energy communities. While Austria has made significant strides in integrating them into its energy landscape, ongoing revisions to the Renewable Energy Expansion Act[6] indicate that further improvements are needed.

This chapter presents a nuanced exploration of an evolving energy paradigm. By shedding light on the motivations, challenges, and opportunities associated with ECs, this research contributes to a deeper understanding of how citizen-led energy initiatives can shape the future of sustainable energy systems. The insights offered in this chapter not only inform ongoing policy discussions but also serve as a valuable resource for practitioners and scholars seeking to navigate the complexities of energy transition in the 21st century.

Urban regions contribute significantly to global greenhouse gas emissions, positioning them as key focal points for sustainability efforts. Cities not only serve as economic and social hubs but also act as substantial energy consumers, necessitating well-structured policies and action plans. The *third chapter* "Reşiţa, Romania – A case study in urban decarbonization

[6]Erneuerbaren Ausbau Gesetz (EAG). EAG 2021. Bundesgesetz über den Ausbau von Energie aus erneuerbaren Quellen (Erneuerbaren-Ausbau-Gesetz - EAG). F.L.G. I 150/2021. 2021.

and climate adaptation through SECAP" is centered on the case study of Reşiţa, Romania, and illustrates the practical application of a comprehensive approach to urban decarbonization and climate adaptation. The Sustainable Energy and Climate Action Plan (SECAP) developed for Reşiţa provides a nuanced perspective on how local governments can align their sustainability strategies with broader national and international goals. By committing to the European Green Deal and the Covenant of Mayors for Climate and Energy, Reşiţa has integrated ambitious climate policies into its governance framework, setting an example for other urban centers across Europe and beyond.

The SECAP initiative in Reşiţa is a multifaceted endeavor that incorporates energy efficiency measures, renewable energy adoption, and resilience-building strategies. Key areas of focus include energy-efficient retrofits for residential and commercial buildings, expansion of photovoltaic energy sources, and transition to sustainable mobility solutions. One of the central aspects of Reşiţa's SECAP is its emphasis on stakeholder engagement. The chapter delves into how collaborative frameworks have facilitated the implementation of SECAP measures, ensuring that diverse perspectives and expertise contribute to the city's decarbonization goals.

Beyond mitigation efforts, Reşiţa's SECAP also prioritizes climate adaptation measures. The chapter explores how risk and vulnerability assessments have informed local strategies to enhance urban resilience. Measures such as green urban spaces, improved water management systems, and infrastructure reinforcements illustrate how cities can proactively prepare for environmental uncertainties. By integrating adaptation strategies alongside mitigation actions, Reşiţa has positioned itself as a model for comprehensive climate governance.

The overarching theme of the chapter aligns with global sustainability efforts. Reşiţa's SECAP directly contributes to SDG 7 (Affordable and Clean Energy), SDG 11 (Sustainable Cities and Communities), and SDG 13 (Climate Action), among others. By contextualizing local action within international frameworks, the case study underscores the importance of multi-level governance in tackling climate challenges. It also highlights how tailored local solutions can serve as blueprints for wider policy implementation at regional, national, and global scales.

The *fourth chapter* "Digital Services for Energy Transition: Enhancing Quality of Life and Delaying Electricity Grid Retrofitting" presents a comprehensive study of how digital solutions can play a transformative role in modern energy landscapes. With a focus on Europe's ambitious energy efficiency targets and the integration of smart technologies, the chapter underscores the necessity of digitalization in optimizing energy consumption, improving grid stability, and fostering consumer engagement.

One of the key themes of this chapter is the potential of digital services to enable a shift from passive energy consumption to active participation by consumers. Through the adoption of digital energy platforms, consumers can become prosumers, actively contributing to energy markets while simultaneously reducing their carbon footprint. The implementation of smart monitoring, automation, and demand-side flexibility is crucial to achieving this transition. By leveraging cutting-edge technologies such as the Internet of Things, artificial intelligence, and blockchain, digital services provide the necessary tools to integrate renewable energy sources seamlessly into existing grids. The InterConnect[7] project, which serves as a central case study in this chapter, exemplifies this transformation through its application of information communication technologies standardization, interoperability, and real-time energy management.

The chapter also highlights the importance of energy efficiency at the community level, with a particular emphasis on positive energy districts and smart urban developments. One such example is the Moneta District in Milan, where digital innovations have been integrated into residential areas to create sustainable living environments. The project demonstrates how energy efficiency measures, such as high-performance insulation, geothermal heating, and smart lighting systems, can collectively reduce energy demand and lower overall emissions. Moreover, the use of digital applications, such as the Planet App, empowers residents to monitor and optimize their energy consumption through user-friendly interfaces and gamified incentives.

Another crucial aspect of the discussion is the role of digital energy services in mitigating the need for costly grid retrofitting. The chapter explores how digitalization can extend the lifespan of existing grids by facilitating more efficient energy distribution, thereby reducing the necessity for large-scale retrofitting projects. Beyond technological advancements, the chapter addresses the policy and regulatory frameworks necessary to support digital energy services as well as the social implications of digital energy services, particularly in the context of smart urban planning.

A key takeaway from the chapter is that energy sustainability is not merely a technological challenge but a systemic one that requires the integration of digital solutions with economic, social, and policy considerations.

[7] https://interconnectproject.eu

The InterConnect project, along with its real-world applications in smart districts like Moneta, provides a blueprint for future energy transitions. By demonstrating the viability of digital energy services in optimizing efficiency, reducing costs, and enhancing grid stability, the chapter highlights the immense potential of digitalization to drive sustainable change.

The advancement of automation systems with energy and resource-saving functions represents one of the most significant trends in modern engineering and sustainability research. In an era where efficiency and sustainability are paramount, the ability to develop intelligent automation systems capable of optimizing energy use and resource management is crucial for both industrial and household applications. The *fifth chapter* "The General Theory of Automation Systems with Energy and Resource Saving Functions" explores the theoretical foundations and practical implementations of such automation systems, demonstrating their impact on contemporary technological and environmental challenges.

This work delves into the concept of energy and resource efficiency, which has emerged as a central concern for societies striving toward sustainable development. This chapter provides a detailed examination of the general theory underlying automation systems that incorporate energy and resource-saving functions, bridging the gap between theoretical control methodologies and practical applications. The foundation of this research lies in the optimization of control processes for both transient and stationary operational modes of automated systems.

The chapter presents a structured approach to defining and solving optimal control problems, which are then transformed into constrained optimization tasks that can be tackled using numerical methods. Such an approach ensures that automation systems operate at peak efficiency while maintaining resource sustainability. This chapter focuses on the application of these theories to specific automated systems, particularly in electric drives and water level controllers. Electric drives, which are widely used in industrial and household applications, serve as an excellent example of how optimal control strategies can significantly improve energy efficiency.

This work is further reinforced by extensive computational simulations conducted using Scilab, an open-source software tool for scientific and engineering applications. The simulations serve to validate the proposed theories, offering concrete evidence of the effectiveness of the methodologies discussed. By leveraging computational modeling, the chapter illustrates how automation systems can be fine-tuned to achieve optimal performance while adhering to energy and resource efficiency principles.

By addressing some important challenges in automation engineering and sustainable development, this work lays the groundwork for future innovations in the field. The integration of optimal control strategies, numerical methods, and computational simulations presents a holistic approach to designing automation systems that not only enhance efficiency but also contribute to a more sustainable future.

The *sixth chapter* "Climate Fitness: Decoding Urban Green Dynamics" highlights how cities can integrate technological innovation, ecological awareness, and social equity to foster sustainable urban environments. As part of the broader discussion presented in "Energy Sustainability and Efficiency: Global Perspectives and Innovative Approaches," this chapter highlights the necessity of redefining urban climate strategies to address the growing complexities of climate adaptation and mitigation.

A key contribution of the chapter is its exploration of real-world applications of climate fitness. Through case studies such as New York's High Line Park and photovoltaic (PV) floor tiles, the research illustrates the benefits and limitations of emerging climate technologies. By applying the climate fitness framework, these case studies provide insights into how urban climate strategies can be refined to ensure equitable and effective sustainability outcomes.

Beyond infrastructure and technology, the chapter also emphasizes the role of community engagement in shaping climate-resilient cities. Public participation in urban planning fosters more inclusive and contextually relevant solutions. Furthermore, the chapter draws attention to the pressing issue of environmental justice. The climate fitness framework, described in the chapter, explicitly incorporates social equity into its assessment criteria, ensuring that sustainability initiatives do not disproportionately benefit affluent populations at the expense of marginalized groups.

The chapter presents a novel perspective on urban sustainability by bridging the gaps between technological innovation, ecological resilience, and social justice. As the challenges posed by climate change continue to intensify, the need for comprehensive and adaptive urban strategies becomes increasingly urgent. This chapter provides a forward-looking framework that equips cities with the tools to navigate these challenges effectively. By embedding the principles of climate fitness into urban planning and energy policies, stakeholders can work toward creating resilient, sustainable, and inclusive urban environments. Within the broader scope of the book, this contribution offers a critical perspective on how cities can harness interdisciplinary insights to drive meaningful and lasting sustainability transformations.

The editors express their deep gratitude to the contributors who provided materials for this book. They would also like to extend their sincere appreciation to the University of Applied Sciences Technikum Wien in general and the Faculty of Industrial Engineering in particular for their invaluable support throughout the preparation of this book.

List of Figures

List of Tables

List of Contributors

Azarmi Khah, Anis, *Planet Smart City, Italy*

Bachleitner, Flora, *University of Applied Sciences Technikum Wien, Austria*

Bonifazi, Alessandro, *Polytechnic University of Bari, Italy*

Ettwein, Frederike, *University of Applied Sciences Technikum Wien, Austria*

Fava, Stefano, *Planet Smart City, Italy*

Grassini, Laura, *Polytechnic University of Bari, Italy*

Helnwein, Jasmin, *University of Applied Sciences Technikum Wien, Austria*

Michel, Mélanie, *University of Applied Sciences Technikum Wien, Austria*

Nevliudov, Igor, *Kharkiv National University of Radio Electronics, Ukraine*

Roccella, Graziella, *Planet Smart City, Italy*

Romashov, Yurii, *Kharkiv National University of Radio Electronics, Ukraine*

Tkachova, Tetiana, *Kharkiv National University of Radio Electronics, Ukraine; XLIM Research Institute, France*

Vaitulevice, Asta, *Ignitis groupe, Lithuania*

Vasilica, Cătălina, *Denkstatt Romania, Romania*

List of Abbreviations

ALM	Actor linkage matrix
BEI	Baseline emission inventory
BEMS	Building energy management systems
BESS	Battery energy storage systems
BLA	Büro für lustige Angelegenheiten
CARES	Community and renewable energy scheme
CA	Climate adaptation
CCTV	Closed-circuit television
CEC	Citizen energy communities
CLISP	Climate change adaptation by spatial planning in the Alpine Space
CNSC	Climate neutral and smart cities
CoM	Covenant of mayors
DER	Distributed energy resources
DSF	Demand side flexibility
DSO	Distribution system operator
DSR	Demand side response
EAG	Erneuerbaren-Ausbau-Gesetz
EC	Energy community
ElWOG	Elektrizitätswirtschafts- und -organisationsgesetz
EMD	Electricity market design / electricity market directive
EMS	Energy management systems
ES	Ecosystem services
EU	European Union
EV	Electric vehicle
GHG	Greenhouse gas
HEMS	Home energy management systems

HLP	High Line Park
IEA	International Energy Agency
ICT	Information communication technologies
IPA	Importance-performance analysis
IPCC	Intergovernmental Panel on Climate Change
IoT	Internet of Things
JPI	Joint Programming Initiative
KPI	Key performance indicator
LCA	Life cycle analysis
MW	Megawatt
NBS	Nature-based solutions
NECP	National energy and climate plans
NGO	Non-governmental organization
NIMBY	Not in my backyard
NZEN	Nearly zero energy neighbourhood
NZN	Net-zero neighbourhood
PAC	Air and climate plan
PED	Positive energy districts
PNRR	National recovery and resilience plan
PV	Photovoltaics
QoL	Quality of life
REC	Renewable energy community
RED	Renewable energy directive
RES	Renewable energy sources
RSE	Research on the energy system
RVA	Risk and vulnerability assessment
SA	Stakeholder analysis
SAREF	Smart applications reference ontology
SDG	Sustainable development goals
SEAP	Sustainability energy action plan
SECAP	Sustainable energy and climate action plan
SET	Strategic energy technology
SGAM	Smart grid architecture model
SNA	Social network analysis
SNE-V	Systemnutzungsentgelte-verordnung
SROI	Social return on investment
TA	Technology assessment

TSO	Transmission system operator
UCR	Urban climate reactor
UGS	Urban green space
UHI	Urban heat island
UHVI	Urban heat vulnerability index
UN	United Nations
VAT	Value added tax

1

Community Engagement for PED Development and Just Energy Transition: A Research Protocol

A. Bonifazi and L. Grassini

Polytechnic University of Bari, Department of Civil, Environmental, Land, Building Engineering and Chemistry (DICATECh), Italy
E-mail: alessandro.bonifazi@poliba.it; laura.grassini@poliba.it

Abstract

Positive energy districts are urban areas that produce more renewable energy than they consume, aiming for net-zero energy imports and zero CO_2 emissions annually. They should bridge the gap between individual building-level energy efficiency and broader city-level climate neutrality goals. Despite constituting a reference terminology for all policies and programs of the European Commission for local-level energy transition, several challenges are still hindering PED's effective operationalization and their contribution to just transitions.

The JPI Urban Europe-funded project Citizens4PED aims to integrate socio-cultural and institutional/regulatory dimensions into the still-evolving technical and economic framework that underpins efforts to develop PEDs. The project particularly highlights the importance of energy justice as a core component of PED development and explores how renewable energy communities can enable just transitions, especially in socially marginalized neighborhoods. For this purpose, Citizens4PED proposes using Living Labs to foster community empowerment in the co-creation of energy solutions and transition pathways and develops a specific research protocol for the analysis of stakeholders and the engagement of the local community, including vulnerable groups. This chapter critically discusses such a protocol, building on stakeholder analysis as a well-established approach to participatory planning

DOI: 10.1201/9788743806851-1

and community engagement, while highlighting its methodological and ethical pitfalls and suggesting potential remedies and alternative research designs.

Keywords: Stakeholder analysis, positive energy districts, renewable energy, public participation, local community.

1.1 Introduction

Cities and urban areas have a huge responsibility in the current climate crisis. With a population share of about 55% on a global scale, which is projected to rise to 70% by 2050 [1], they consume 75% of primary energy and generate about 70% of global CO_2-eq emissions [2]. Because of the high exposure of their infrastructure, people, and businesses, they are also highly vulnerable to climate change effects.[8] On the other side, thanks to the concentration of human and financial capital, cities and urban areas may also play a leadership role in addressing climate change and become key agents for change toward energy transition [3]. This frontline role has been increasingly acknowledged in international agreements for sustainable development like the Agenda 2030 [4], the New Urban Agenda [5], and the European Green Deal [6].

The concept of positive energy districts (PEDs) is strongly linked to the dual role played by urban areas and cities in the current climate and energy crisis. It refers to a district with annual net zero energy import and net zero CO_2 emissions, working toward an annual local surplus production of renewable energy.[9] Firstly introduced in the revision of the Strategic Energy Technology (SET) Plan of the European Union as a key pillar for the Implementation Plan of its Sub-Action 3.2 "Smart Cities and Communities" [7–9], it now constitutes a reference terminology for all policies and programs of the European Commission for local-level energy transition and a cornerstone in the 2030 Climate Neutral and Smart Cities (CNSC) Mission of the European Union [10].

---

[8]Since the Fifth IPCC report (IPCC, 2014), a specific focus on urban areas was also included in the Assessment Reports.

[9]The SET Plan introduces the concept as follows: "Positive energy districts are energy-efficient and energy-flexible urban areas or groups of connected buildings which produce net zero greenhouse gas emissions and actively manage an annual local or regional surplus production of renewable energy. They require integration of different systems and infrastructures and interaction between buildings, the users and the regional energy, mobility and ICT systems, while securing the energy supply and a good life for all in line with social, economic and environmental sustainability" (JPI Urban Europe, 2020a).

Compared to the sister concepts of Net-Zero Neighbourhood (NZN) and Nearly Zero Energy Neighbourhood (NZEN), it places a large emphasis on the renewable energy production pillar besides the energy efficiency one (which constitutes the core concern of the latter) and the energy flexibility. Its introduction in the debate on the energy transition of cities had also the merit of underlining the importance of the neighborhood scale in the process, by bridging the gap between the efforts traditionally placed on energy efficiency at the building scale and the wider focus on the city level [11, 12].

Despite the broad definition of PEDs, research to date has mainly focused on technical issues related to renewable energy sources development and efficiency, while little attention has been paid to social issues and the distributional effects of PED transition[10] [13]. The social aspects of PEDs are addressed in a relatively small number of papers, mainly dealing with social innovation related to energy citizenship [14] and stakeholder engagement [15–17], with an emphasis on public acceptance and behavioral changes connected to energy uses [18]. Few papers explicitly highlight the need to focus on social justice concepts in PED development [19] and to adopt a "human-centric approach", as a bottom-up perspective prioritizing human and ecosystem well-being and the lived experiences [20]; very few underline the challenges of energy access and energy poverty [19–21].

The JPI Urban Europe funded project Citizens4PED – on which this chapter is based – explicitly aims to contribute to the PED debate by integrating the socio-cultural and institutional/regulatory dimensions with the techno-energetic one. It deals with PED transition in four existing neighborhoods in three countries (Belgium, Italy and Austria) and explores synergies with renewable energy communities as key enabling mechanisms for just transitions. In so doing, it takes further the initial research made on the potential of the concept of energy communities for PED development, as a means to enhance social innovation and tackle energy poverty [15, 17].

Citizens4PED particularly emphasizes the energy justice perspective, as not only is energy poverty unevenly distributed across the EU – both socially and spatially – but transition policies may unexpectedly create further inequalities according to the multiple dimensions of energy justice [19, 22]. These include the distributional dimension, which is linked to the fair distribution of benefits and burdens [23, 24]; the recognitional dimension, which is

[10]This is despite the explicit mention, in the PED definition (see previous note), of the need to secure energy supply and a good life for all in line with the principles of social sustainability.

related to the recognition of vulnerabilities and different needs, as well as to respect for diversity [25, 26]; and the procedural dimension, which is referred to equitable access to information and decision-making [27].

Because of this perspective, the overall research design of Citizens4PED aims to increase knowledge about which processes can promote just transitions toward PEDs in different urban contexts, including very challenging neighborhoods such as social housing estates, where problems of physical degradation of the building stocks may be combined with low accessibility, socioeconomic marginalization, and social stigma.

In those contexts, Citizens4PED envisages the development of Living Labs as reflexive, adaptive, and multi-actor learning environments, where knowledge and solutions may be co-created and tested within real-world contexts [28] and just local energy transition partnerships built, thus paving the way to community empowerment and transformative alliances for energy transition at wider scales [29]. From an energy justice perspective, a core concern is the stakeholder engagement methodology and its capacity to empower even the most vulnerable groups, who are often chronically hard to reach. This chapter deals with the development of a research protocol for this purpose, based on a critical assessment of a widespread and yet diversified (and often contentious) approach: stakeholder analysis (SA).

The chapter's outline includes an introduction to SA (Section 2) followed by a description of the proposed research protocol organized in five steps (Section 3) and some concluding reflections (Section 4).

1.2 Stakeholder Analysis

Stakeholder analysis is a well-established set of methods for studying socio-ecological systems by focusing on key actors and by assessing their respective interests and ability to influence actual or potential change [30]. Stakeholder analysis can be defined as an approach to generating knowledge about actors – both individuals and organizations – by analyzing their intentions and relationships and assessing the resources and capabilities they bring to bear on decision-making or implementation processes [31]. In practical terms, SA aims to [32]:

- Identify, characterize, or categorize the stakeholders who can influence and possibly change an organization, process, or system.
- Develop an understanding of why change occurs in that organization, process, or system.

- Understand who can make changes happen (both in terms of individual actors and networks).
- Discuss how best to manage, for example, the local energy transition in a particular district.

Stakeholder analysis has been widely used in environmental and sustainability research, including energy transition studies [33]. However, many scholars and practitioners warn of the risk of SA supporting oversimplified, manipulative, and patronizing views of social networks, not least because the definition of stakeholder itself remains controversial [32, 34]. In this respect, we value the opportunity to engage unrepresented subjects (non-stakeholders), while being wary of the risk of empowering already dominant stakeholders [35–36].

Despite its limitations, and provided that appropriate methodological checks and balances are fully implemented, a Community Engagement (CE) strategy based on SA can help advance the living lab activities in each target area in three important directions [37]:

- CE can initiate dialogue and facilitate discussions between previously unconnected actors while revealing power asymmetries between stakeholders.
- When interests are explicitly addressed, CE can facilitate a common understanding of key issues and help strengthen cooperation, while also highlighting incommensurable or irreconcilable views.
- In particular, critical CE can enhance social responsibility and encourage organizations and individuals to change their practices in response to public participation.

In the following section, the reflection on SA as an effective approach to foster community engagement in local energy transitions develops along the steps of the research protocol that has been implemented in the framework of the Citizens4PEDs project. The content's layout notwithstanding, ample room is earmarked to discuss the broader implications of the methodology for energy governance and urban policymaking.

1.3 A Community Engagement protocol for Citizens4PED

The proposed protocol for CE unfolds in five successive steps. While different methods and tools may be used according to local expertise and the specificities of the study contexts, an emphasis is put on critical approaches, based on participatory action-research methods [38].

1.3.1 Scoping the PED study area

To identify the study area in PED development processes, it might be advisable to start with a specific target area or neighborhood, and then reach out to other parts of the district/municipality following relevant socio-spatial relationships. To fully capture the interactions between energy practices and PED-relevant policies and regulations, it may also be necessary to cover a whole city or region. On a practical level, the investigation of social networks could be centered on the people living in, and on the organizations located in the target area, but may involve actors at different levels (thus including, among others, metropolitan government, distribution network operator).

1.3.2 Stakeholder identification

The identification of relevant stakeholders can be carried out using a mix of possible methods (including document analysis and participant observation, as well as semi-structured interviews, focus groups, and other intersubjective research methods that allow for snowball sampling). The outcome can be mapped on a matrix linking a tentative list of types of stakeholders (local authorities, social housing management agencies, residents, energy providers, etc.) with potential roles in PED development (producer, consumer, prosumer, facilitator, investor, etc.) while also providing a tentative description of motivations and interests for each stakeholder [39]. In line with the methodological assumptions adopted in this study, motivations and interests can be used, inductively, to reach out to uninvolved, underrepresented stakeholders, starting with those who may be at risk of energy poverty or who lack a voice in decision-making processes or suffer from social exclusion dynamics [40–41].

1.3.3 Stakeholder characterization and categorization

Stakeholder characterization and categorization builds on the previous step– in which each subject was described in terms of organizational type, potential role in PED development, motivations, and interests – to further explore the crucial dimensions that are likely to affect the role of stakeholders in the PED-development process and, more importantly, the current and future dynamics of social networking around local energy transitions. Luyet and coworkers [36] reviewed alternative approaches to stakeholder characterization, according to:

- Attitudes toward or interest in, a project or process;
- Potential conflicts and coalitions between stakeholders and objectives;

- Access to resources;
- Political influence on the project;
- Degree of implication or commitment;
- Power, urgency, proximity, and legitimacy.

Furthermore, the step can either be carried out in an *analytical* (top-down) manner by researchers alone, or it can involve stakeholders themselves in a so-called *reconstructive* (bottom-up) categorization [34]. Finally, unlike in the primary streams of research in the stakeholder theory tradition [41], the focus of Citizens4PED is not on supporting public authorities or private companies in managing their relationships with, or moral obligations toward, local actors and networks. Rather, the research protocol aims to empower the innovative, multi-actor PED coalitions emerging within the Living Labs.

Consequently, the research protocol at this stage can be summarized as follows:

- Stakeholder characterization and categorization should always be carried out collaboratively by researchers, involving as much as possible the stakeholders themselves (through questionnaires, interviews, etc).
- Stakeholders should be provisionally characterized, at the time the research is carried out and taking into account the dependency of the exercise on the actual participants and contextual conditions, according to three dimensions:
    - Resources (material, financial, etc.)
    - Capabilities (management, technological, relational, cultural, etc.)
    - Engagement in PED development.

A preliminary categorization can be performed as presented in Figure 1.1 for the sole purpose of triggering further investigation [43], while a revised categorization model may emerge as a project outcome.

As the terms employed in the characterization and categorization model have multiple meanings, it is important to signal that, in the context of this research protocol, they are to be understood primarily as they are in the *capabilities framework* originally developed by Amartya Sen and Martha Nussbaum [44] and in the *resource-based view* approach to socioeconomic innovation [45].

1.3.4 Reconstructing stakeholder networks

One mandatory and three optional methods are proposed to reconstruct stakeholder interactions. The Actor Linkage Matrix (ALM) should be the basic

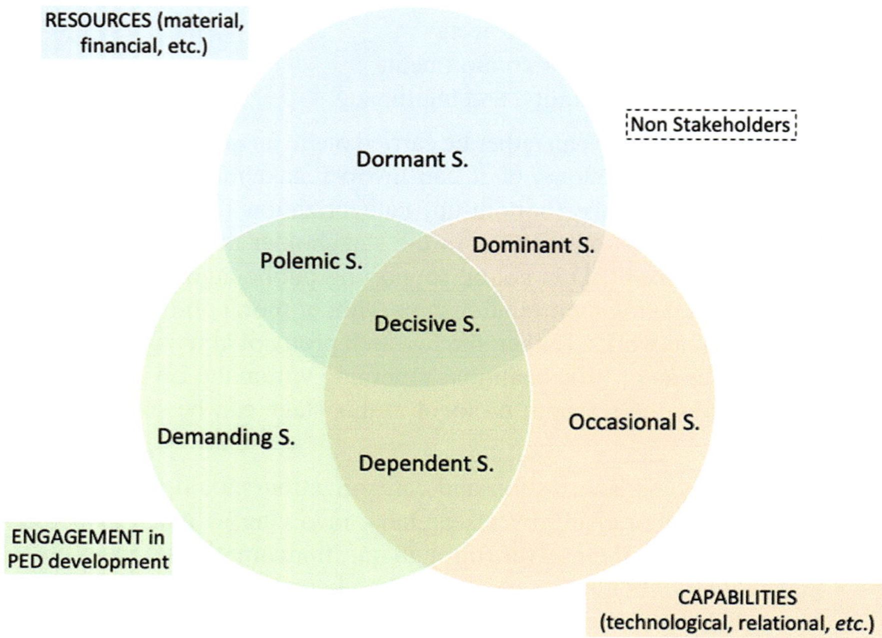

Figure 1.1 A tentative model for stakeholder characterization and categorization. Adapted from [43].

tool used in all target areas to collect data on relationships between local stakeholders. It can be used to map the existing situation and to highlight key linkages, but it can also provide a rationale for transformative action [46–48].

The ALM is a square matrix with the same stakeholders in both rows and columns, and the cells represent the interactions between each pair of stakeholders: in the framework of the Citizens4PED project, each cell was used to summarize the degree of interaction according to a 7-class ordinal scale ranging from "high conflict" to "high cooperation", through the corresponding medium and low classes and the "no interaction" case, as shown in Table 1.1. Building on the bottom-up approach adopted, the ALM can be filled in directly by the stakeholders or by the researchers on the basis of an interview or the administration of a mini-questionnaire. In this case, we use "stated measures" of stakeholder interaction, but "derived measures" may also be used, by tapping into participant observation, document analysis, or various primary or secondary data collection [49]. Finally, researchers may mix different analytical techniques and research designs (bottom-up vs. top-down).

In Table 1.1, we simulate the result of a research design based on stated measures that also cover the directionality of the interaction, i.e., two stated measures of conflict or cooperation according to each stakeholder in a pair are shown in a different cell. For example, the fourth cell in the second row shows how representatives of the local authority perceive their level of cooperation with the community enterprise. In contrast, the second cell in the fourth row reflects how members of the community enterprise assess their cooperation with the local authority–assumed, in this case, to differ.

However, the discussion of reciprocity should be limited to varying degrees of either conflict or cooperation, as ambivalent relationships would require more sophisticated analytical tools [50]. A simplified version of the ALM would omit the examination of reciprocity and only fill in the lower half of the matrix, below the diagonal.

The accompanying text could be used, for each pair of stakeholders, to provide a better insight into their interaction. Although the issue deserves further elaboration, it should be noted that both organizations (e.g., the municipal council or the distribution system operator) and collective actors (e.g., residents, energy managers) should be modeled as individual stakeholders; however, if there is evidence of PED-relevant differentiation within such entities, sub-groups could be identified that may partially overlap with the

Table 1.1 Actor Linkage Matrix for a PED development process, based on [46–48]. An example of reciprocity is highlighted by underlining the corresponding stated measures by each stakeholder in a pair.

	A) Local authority	**B) Energy provider**	**C) Community enterprise**	**D) Residents**
A) Local authority		Low cooperation	Medium cooperation	Low conflict
B) Energy provider	Low cooperation		No interaction	Low conflict
C) Community enterprise	High cooperation	No interaction		Medium cooperation
D) Residents	High conflict	Low conflict	High cooperation	

parent category (the opposition coalition, the residents of the High Street, etc.). Unless they can be identified as actors playing relatively autonomous roles (e.g., local leaders) who need to be analyzed as such, individuals who straddle different categories are expected to be subsumed under each organization or group to which they belong.

Furthermore, ALM can be used either as a standalone tool or to feed relational data to a more sophisticated research method such as social network analysis (SNA), which aims to describe the composition of, and interaction in, a network, by measuring the web of connections between agents or entities, and the importance of each relative to the network as a whole or to any grouping of entities [51]. Accordingly, PED networks are defined by organizations or stakeholders ("nodes") connected by links ("edges"), based on existing (secondary) datasets on any kind of interaction (including previous collaborative sustainability or local development projects), or drawing from the ALM alone. In the latter case, the most meaningful information would come from the mere occurrence of links between certain stakeholders, although the strength of the cooperation (or conflict) could be used to weigh the edges. A set of indicators on the intensity of links and the configuration of the network can be calculated to identify the nodes or edges that play a more strategic role [52].

Energy community-building applications of SNA are well established [53] and the open-source software package Gephi Tools[11] is most often used both to measure the indicators and to draw the network diagrams [54]. An application of SNA showing the output of a similar software (iGraph R package) and focusing on adaptation in socio-ecological systems is shown in Figure 1.2.

Although they could complement formal SNA, concept mapping software such as Obsidian (https://obsidian.md) or other knowledge management tools, such as Kumu (https://kumu.io), could also be the main stakeholder network mapping tools to track contacts and sketch networks, while linking notes, images, documents, videos, audio, and web pages to the different stakeholders. This could prove useful for both research management objectives and local dissemination and engagement efforts. Figure 1.3 shows one of the views created with Kumu under the Making-City's PED-focused Horizon 2020 research project, to map the interconnections between the Smart Cities and Communities Lighthouse projects' stakeholders.

[11]https://gephi.org.

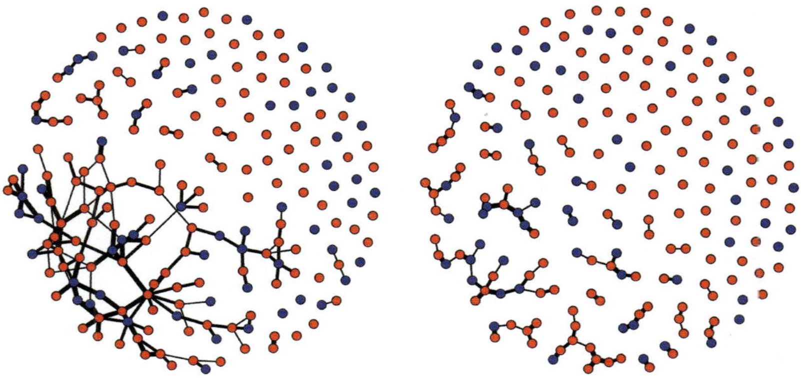

Figure 1.2 Using Social Network Analysis to model peer influence networks based on actor similarity (left) or random interactions (right). Adapted from [55].

Provided that steps 1–3 are completed, and the Actor Linkage Matrix is filled in, researchers and the involved stakeholders might want to opt out of SNA and choose other methods to complete their investigation of social dynamics in the target areas – as diverse as agent-based modeling [56] and community autoethnography [57–58].

1.3.5 Networking and PED-community building

The protocol concludes with a brief reminder that stakeholder analysis in the framework of the Citizens4PED project and of other PED-related participatory processes should be conducted in an open and transparent process, with the primary aim of empowering local stakeholders to build just local energy transition partnerships. To this end, CE can be informed by different conceptual approaches and research designs [60]. The dynamic role of stakeholders should always be duly taken into account, as they may hinder or promote PED development processes, also depending on the expected distribution of benefits over time, while they seem to respond to stable policy support frameworks [50].

As project activities progress, additional methodologies are likely to complement the proposed Community Engagement protocol. These may focus on social networks as units of analysis and be more attuned to resource sharing, collective capability building, and the dynamics of mutual engagement, with the aim of fully empowering local stakeholders [38].

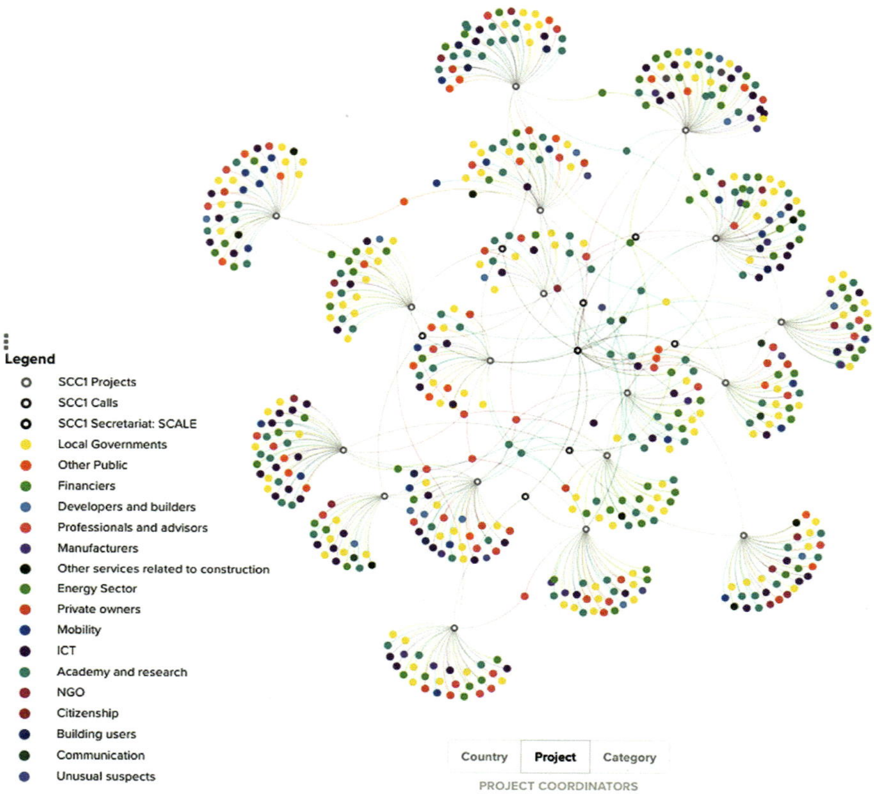

Figure 1.3 A Smart Cities and Communities Lighthouse projects stakeholders' map, developed with the Kumu tool in the framework of the Making-City Horizon 2020 project (https://makingcity.eu). Adapted from [59].

1.4 Concluding Remarks

Community-led approaches to energy transition are crucial for just transitions. They may empower local stakeholders through local cooperation in the face of power imbalances and the commodification of energy [61]. These initiatives can support just local transition coalitions and foster new opportunities for socioeconomic development at the local level, while enhancing energy security and contrasting energy poverty [62], thus contributing to the right to energy by vulnerable groups.

In this Chapter, we have discussed a research protocol for Community Engagement that builds on the analysis of stakeholders, including their potential roles, relationships, and influences in PED development. It was elaborated within the JPI Urban Europe-funded project Citizens4PED, which specifically aims to contribute to enriching the knowledge about which processes can promote just transitions toward PEDs in different urban contexts, including very challenging neighborhoods. The protocol has been applied in the initial phase of the Living Labs established by the project in some target areas in Bari (Italy), Brussels (Belgium) and Vienna (Austria), with the final aim to empower local stakeholders to co-construct energy solutions and just local transition pathways.

As the Living Labs are still in the initial phase of development, a full assessment of the results of the CE application cannot be made, yet. Nevertheless, the application of the protocol already proved very useful: to initiate dialogue and facilitate discussions; to highlight common understanding and irreconcilable views; to highlight power dynamics, motivations, and potential for collaboration; to device the most appropriate engagement and management strategies for the follow-up co-creation phase of the Living Labs.

Acknowledgments

This article is based on the research carried out within the project Citizens4PED (2022–2025), which has been funded under the JPI-Urban Europe Call 2021 for Positive energy districts and Neighbourhoods for Climate Neutrality.

References

[1] United Nations, "The Sustainable Development Goals Report 2024," 2024. [Online]. Available: https://unstats.un.org/sdgs/report/2024/.

[2] IPCC, "Climate Change 2022: Mitigation of Climate Change. Working Group III Contribution to the Sixth Assessment Report of the Intergovernmental Panel on Climate Change," Cambridge University Press, Cambridge, UK, 2021.

[3] D. Reckien, M. Salvia, O. Heidrich, J. M. Church, ... and R. Dawson, 'How are cities planning to respond to climate change? Assessment of local climate plans from 885 cities in the EU-28,' Journal of Cleaner Production, vol. 191, pp. 207–219, 2018.

[4] United Nations, 'Transforming our world: the 2030 Agenda for Sustainable Development,' 2015.

[5] United Nations, 'New Urban Agenda,' The United Nations publication, Habitat III Secretariat, A/RES/71/256, 2017.
[6] European Commission, 'Decarbonising our energy system to meet our climate goals,' European Commission, 2021.
[7] JPI Urban Europe, 'Implementation Plan: Europe to become a global role model in integrated, innovative solutions for the planning, deployment, and replication of Positive Energy Districts,' JPI Urban Europe and JPI Urban Europe / SET-Plan Action 3.2, 2018.
[8] JPI Urban Europe, 'White Paper on PED Reference Framework for Positive Energy Districts and Neighbourhoods,' JPI Urban Europe and SET-Plan Action 3.2, 2020.
[9] JPI Urban Europe, 'Europe towards positive energy districts: A compilation of projects towards sustainable urbanization and the energy transition,' JPI Urban Europe / SET-Plan Action 3.2, 2020.
[10] European Commission, 'EU Mission: Climate-Neutral and Smart Cities,' European Commission, 2021.
[11] M. Amado, F. Poggi, and A. R. Amado, 'Energy efficient city: a model for urban planning,' Sustainable Cities and Societies, vol. 26, pp. 476–485, 2016.
[12] N. Komninos, 'Net zero energy districts: Connected intelligence for carbon-neutral cities,' Land, vol. 11, p. 210, 2022.
[13] L. N. Sassenou, L. Olivieri, and F. Olivieri, 'Challenges for positive energy districts deployment: A systematic review,' Renewable and Sustainable Energy Reviews, vol. 191, p. 114152, 2024.
[14] M. van Wees, B. P. Revilla, H. Fitzgerald, D. Ahlers, N. Romero, ... and S. Smit, 'Energy citizenship in positive energy districts. Towards a transdisciplinary approach to impact assessment,' Buildings, vol. 12, p. 186, 2022.
[15] A. Tuerk, D. Frieden, C. Neumann, K. Latanis, A. Tsitsanis, ... and T. Ramschak, 'Integrating plus energy buildings and districts with the EU energy community framework: regulatory opportunities, barriers and technological solutions,' Buildings, vol. 11, no. 10, p. 468, 2021.
[16] D. Baer, B. Loewen, C. Cheng, J. Thomsen, A. Wyckmans, ... and D. Ahlers, 'Approaches to social innovation in positive energy districts (PEDs). A comparison of Norwegian projects,' Sustainability, vol. 13, no. 13, p. 7362, 2021.
[17] D. Mihailova, I. Schubert, P. Burger, and M. M. C. Fritz, 'Exploring modes of sustainable value co-creation in renewable energy communities,' Journal of Cleaner Production, vol. 330, p. 129917, 2022.

[18] S. A. Petersen, I. Petersen, and P. Ahcin, 'Smiling earth. Raising awareness among citizens for behaviour change to reduce carbon footprint,' Energies, vol. 13, p. 5932, 2020.
[19] A. X. Hearn, A. Sohre, and P. Burger, 'Innovative but unjust? Analysing the opportunities and justice issues within positive energy districts in Europe,' Energy Research & Social Science, vol. 78, p. 102127, 2021.
[20] M. T. Nguyen and S. Batel, 'A critical framework to develop human-centric positive energy districts: Towards justice, inclusion, and well-being,' Frontiers in Sustainable Cities, vol. 3, p. 691236, 2021.
[21] A. Nordholm and S. Sareen, 'Scalar containment of energy justice and its democratic discontents: Solar power and energy poverty alleviation,' Frontiers in Sustainable Cities, vol. 3, p. 13, 2021.
[22] K. Jenkins, B. K. Sovacool, and D. McCauley, 'Humanizing sociotechnical transitions through energy justice: An ethical framework for global transformative change,' Energy Policy, vol. 117, pp. 66–74, 2018.
[23] F. Bartiaux, M. Maretti, A. Cartone, P. Biermann, and V. Krasteva, 'Sustainable energy transitions and social inequalities in energy access: A relational comparison of capabilities in three European countries,' Global Transitions, vol. 1, pp. 226–240, 2019.
[24] C. Groves, F. Shirani, N. Pidgeon, C. Cherry, G. Thomas, ... and K. Henwood, 'The bills are a brick wall': Narratives of energy vulnerability, poverty and adaptation in South Wales,' Energy Research & Social Science, vol. 70, p. 101777, 2020.
[25] K. Jenkins, D. McCauley, R. Heffron, H. Stephan, and R. Rehner, 'Energy justice: A conceptual review,' Energy Research & Social Science, vol. 11, pp. 174–182, 2016.
[26] R. Gillard, C. Snell, M. Bevan, 'Advancing an energy justice perspective of fuel poverty: Household vulnerability and domestic retrofit policy in the United Kingdom', Energy Research & Social Science, vol. 29, pp. 53-61, 2017.
[27] S. Shejale, M.X. Zhan, M. Sahakian, R. Aleksieva, M.E. Biresselioglu,... and K.L. Schibel, 'Participation as a pathway to procedural justice: A review of energy initiatives across eight European countries', Energy Research & Social Science, 122, 103982, 2025.
[28] E. Puerari, J. de Koning, T. Von Wirth, P. Karré, I. Mulder, D. Loorbach, 'Co-Creation dynamics in urban living labs', Sustainability, vol. 10, no. 6, p. 1893, 2018.

[29] K. Hölscher, N. Frantzeskaki (Eds.), 'Transformative Climate Governance: A Capacities Perspective to Systematize, Evaluate and Guide Climate Action', Palgrave Macmillan, London, UK, 2020.
[30] R. Grimble, K. Wellard, 'Stakeholder methodologies in natural resource management: a review of principles, contexts, experiences and opportunities', Agricultural Systems, vol. 55, no. 2, pp. 173–193, 1997.
[31] Z. Varvasovszky, R. Brugha, 'How to do (or not to do)... A stakeholder analysis', Health Policy and Planning, vol. 15, no. 3, pp. 338–345, 2000.
[32] C. Billgren, H. Holmén, 'Approaching reality: Comparing stakeholder analysis and cultural theory in the context of natural resource management', Land Use Policy, vol. 25, no. 4, pp. 550–562, 2008.
[33] Z. D. W. Putra, W. van der Knaap, 'A smart city needs more than just technology: Amsterdam's energy atlas project', in: L. Anthopoulos (Ed.), Smart City Emergence, Elsevier, Amsterdam, pp. 129–147, 2019.
[34] M. S. Reed, A. Graves, N. Dandy, H. Posthumus, K. Hubacek, J. Morris, C. Prell, C. H. Quinn, L. C. Stringer, 'Who's in and Why? A Typology of Stakeholder Analysis Methods for Natural Resource Management', Journal of Environmental Management, vol. 90, no. 5, pp. 1933–1949, 2009.
[35] B. Junker, M. Buchecker, U. Müller-Böker, 'Objectives of public participation: Which actors should be involved in the decision making for river restorations?', Water Resources Research, vol. 43, no. 10, p. 10438, 2007.
[36] V. Luyet, R. Schlaepfer, M. B. Parlange, A. Buttler, 'A framework to implement stakeholder participation in environmental projects', Journal of Environmental Management, vol. 111, pp. 213–219, 2012.
[37] B. K. Sovacool, 'A critical stakeholder analysis of the Trans-ASEAN Gas Pipeline (TAGP) Network', Land Use Policy, vol. 27, no. 3, pp. 788–797, 2010.
[38] L. M. Vaughn, F. Jacquez, 'Participatory Research Methods – Choice Points in the Research Process', Journal of Participatory Research Methods, vol. 1, no. 1, 2020.
[39] P. Selman, 'Community participation in the planning and management of cultural landscapes', Journal of Environmental Planning and Management, vol. 47, no. 3, pp. 365–392, 2004.
[40] K. Grossmann, 'Using conflicts to uncover injustices in energy transitions: The case of social impacts of energy efficiency policies in the housing sector in Germany', Global Transitions, vol. 1, pp. 148–156, 2019.

[41] J. Hills, 'Fuel poverty-the problem and its measurement: Interim report of the Fuel Poverty Review', CASE report, vol. 69, Centre for Analysis of Social Exclusion, London School of Economics and Political Science, London, UK, 2011.
[42] K. D. Butterfield, R. Reed, D. J. Lemak, 'An Inductive Model of Collaboration - From the Stakeholder's Perspective', Business & Society, vol. 43, no. 2, pp. 162–195, 2004.
[43] R. K. Mitchell, B. R. Agle, D. J. Wood, 'Toward a Theory of Stakeholder Identification and Salience: Defining the Principle of Who and What Really Counts', The Academy of Management Review, vol. 22, no. 4, p. 853, 1997.
[44] R. Day, G. Walker, N. Simcock, 'Conceptualising energy use and energy poverty using a capabilities framework', Energy Policy, vol. 93, pp. 255–264, 2016.
[45] S. Khanra, P. Kaur, R. P. Joseph, A. Malik, A. Dhir, 'A resource-based view of green innovation as a strategic firm resource: Present status and future directions', Business Strategy and the Environment, vol. 31, no. 4, pp. 1395–1413, 2022.
[46] S. Biggs, H. Matsaert, 'An actor-oriented approach for strengthening research and development capabilities in natural resource systems', Public Administration and Development, vol. 19, pp. 231–262, 1999.
[47] S. Biggs, H. Matsaert, 'Strengthening poverty reduction programmes using an actor-oriented approach: examples from natural resources innovation systems', AgREN Network Paper, vol. 134, Overseas Development Institute, London, UK, 2004.
[48] B. Suchiradipta, S. Raj, 'Agricultural Innovation Systems (AIS): A Study of Stakeholders and their Relations in System of Rice Intensification (SRI)', The Journal of Agricultural Education and Extension, vol. 21, no. 4, pp. 343–368, 2015.
[49] S. J. Greenland, I. A. Combe, A. M. Farrell, 'Stakeholder Preference and Stated vs Derived Importance Satisfaction Research', International Journal of Market Research, vol. 58, no. 1, pp. 35–56, 2016.
[50] S. Ruggiero, T. Onkila, V. Kuittinen, 'Realizing the social acceptance of community renewable energy: A process-outcome analysis of stakeholder influence', Energy Research and Social Science, vol. 4, pp. 53–63, 2014.
[51] S. G. Enger, M. Gulbrandsen, 'Orchestrating collaborative projects: inside ICT networks in Horizon 2020', Science and Public Policy, vol. 47, no. 3, pp. 396–409, 2020.

[52] J. Scott, 'Social Network Analysis', Sage, London, UK, 2012.
[53] F. Goedkoop, J. Dijkstra, A. Flache, 'A social network perspective on involvement in community energy initiatives: The role of direct and extended social ties to initiators', Energy Policy, vol. 171, p. 113260, 2022.
[54] M. Bastian, S. Heymann, M. Jacomy, 'Gephi: an open-source software for exploring and manipulating networks', in: Third International AAAI Conference on Weblogs and Social Media, 2009.
[55] J. Yletyinen, G. L. W. Perry, P. Stahlmann-Brown, R. Pech, J. M. Tylianakis, 'Multiple social network influences can generate unexpected environmental outcomes', Scientific Reports, vol. 11, p. 9768, 2021.
[56] A. Akhatova, L. Kranzl, F. Schipfer, C. B. Heendeniya, 'Agent-Based Modelling of Urban District Energy System Decarbonisation—A Systematic Literature Review', Energies, vol. 15, no. 2, p. 554, 2022.
[57] M. Haeffner, F. Hames, M. M. Barbour, J. M. Reeves, G. Platell, S. Grover, 'Expanding collaborative autoethnography into the world of natural science for transdisciplinary teams', One Earth, vol. 5, no. 2, pp. 157–167, 2022.
[58] M. Sarrica, M. Richter, S. Thomas, I. Graham, B. M. Mazzara, 'Social approaches to energy transition cases in rural Italy, Indonesia and Australia: Iterative methodologies and participatory epistemologies', Energy Research & Social Science, vol. 45, pp. 287–296, 2018.
[59] Making-City, 'SCC1 Projects Stakeholders' Map', available at: https://embed.kumu.io/ebcd39c318a26e579861ae0e52dc673e?settings=0#ssc1-projects-stakeholders-map [accessed Dec. 4, 2024].
[60] G. A. Marcon Nora, A. Alberton, D. H. F. Ayala, 'Stakeholder theory and actor-network theory: The stakeholder engagement in energy transitions', Business Strategy and the Environment, vol. 32, no. 1, pp. 673–685, 2023.
[61] A. M. Feldpausch-Parker, D. Endres, T. R. Peterson, S. L. Gomez (Eds.), 'Routledge handbook of energy democracy', 1st ed., Routledge, London, 2021.
[62] P. Kivimaa, W. P. Boon, G. P. J. Verbong, 'Energy transition and the role of communities', Energy Research & Social Science, vol. 50, pp. 1–10, 2019.

2

Revolution or Illusion? Austrian Visions and Realities of Energy Communities

F. Ettwein, M. Michel, and F. Bachleitner

University of Applied Sciences Technikum Wien, Austria
E-mail: melanie.michel@technikum-wien.at;
flora.bachleitner@technikum-wien.at

Abstract

Since the Austrian Renewable Energy Expansion Act came into force in 2021, approximately 3.400 energy communities (ECs) have been registered in Austria. However, it remains to be seen whether this upward trend will continue, highlighting the need to gain a deeper understanding of citizens' expectations and perceptions of ECs to further promote their adoption. Therefore, this study explores whether ECs meet citizens' expectations and needs. An importance-performance analysis with a sample of 1,073 Austrian residents was conducted. The results indicate that financial and environmental aspects are fundamental to potential EC members. On the other hand, social aspects are seen as less important and, thus, not decisive factors for joining an EC. Concerns exist regarding supply security and data security within ECs. This is linked to a general lack of information and public knowledge, which can result in the spread of incorrect assumptions about the functioning and values of ECs. To address these concerns, it is crucial to launch rigorous information and education campaigns that strengthen public dialogue and understanding of ECs.

Keywords: Renewable energy community, perception, motives and challenges, preconception, importance-performance analysis.

DOI: 10.1201/9788743806851-2

2.1 Introduction

With the Clean Energy for All Europeans Package, the EU has set a path for a sustainable energy transition with ambitious targets for the expansion of renewables. One initiative aimed at these targets is the establishment of energy communities (ECs) as laid down in the Renewables Energy Directive (RED II and RED III) [1, 2] and the Electricity Market Directive (EMD) [3]. Austria transposed these legal requirements into national law through the Renewable Energy Expansion Act (Erneuerbaren Ausbau Gesetz, EAG) [4]. Since these laws were enacted in 2021, so-called renewable energy communities (REC) and citizen energy communities (CEC) have a regulatory base in Austria. With the establishment of ECs, citizens are placed at the center of the energy transition, evolving from passive consumers to active prosumers who shape their community's power supply – a narrative frequently highlighted in both legislative and scientific literature [5–9]. Advisory offices and service providers for ECs incorporate this narrative and advertise their services and community memberships with various benefits for citizens. These benefits not only include the democratization of energy supply but also social, technical, environmental, and economic advantages [10–18] (which we dive deeper into in section 2). The same applies to academic pilot projects[12] researching ECs to shape them based on citizen's needs. The common goal is to motivate citizens to join or even find their own EC and establish them as a new energy supply standard in Austria. The question remains: Are these strategies effective? Do energy service providers send the right messages to citizens? Furthermore, do pilot communities meet citizens' needs?

At present state, about three years after the EAG entered into force, more than 3.400 ECs have been officially registered in Austria [19] (query from July 2024), and there has been comprehensive media coverage on the topic. However, a mere registration does not indicate actual operation, so the numbers alone are unsuitable for evaluating the success of energy communities in Austria. Still, studies indicate a substantial level of interest in ECs among the Austrian population: a representative survey has shown a high level of interest in ECs among the Austrian public, with about two-thirds of people expressing interest in joining an EC [20]. This survey represents a promising starting point for future developments of ECs in Austria, as this suggests a high potential for future community growth. To maintain this upward trend

[12]There are several European research projects dealing, amongst others, with citizens' needs toward community energy, e.g., Compile, Wisegrid, GRETA, SHAREs; ENCLUDE etc.

and reach a wider audience, it is vital to understand citizens' expectations and impressions of ECs. Only then can communication strategies effectively provide the correct information on the benefits of ECs to their customers and fully leverage their potential. While there is research on the motives for joining an EC and the challenges their members face (see Section 2), it remains unclear whether EC initiatives can effectively address these factors.

Further, the results allow us to determine misconceptions among residents: as ECs are still quite a new and innovative concept to most, there are misguided assumptions about the reality of an EC. Joining an EC with inaccurate expectations can lead to disappointment and frustration among the members, ultimately hindering citizens' willingness to participate in ECs and blocking the full potential of ECs. Additionally, wrong expectations may lead to a refusal of EC membership. At the heart of this lies the following research question: To what extent do energy communities align with citizens' expectations in Austria, and which barriers arise as a result?

To address this knowledge gap, this study gathered data from 1,073 Austrian citizens, focusing on their general expectations regarding (renewable) ECs – whether inflated or realistic – and perceptions of existing participation barriers. Using an importance-performance analysis, the study examines whether citizens' expectations are met and if the proposed values align with the needs of future EC members. This citizen-centric approach aims to uncover the potential disconnect between the advertised benefits of ECs and the motivations driving individuals to join an EC or find their own EC.

The remainder of this paper is structured as follows: Section 2 provides the theoretical foundation on ECs in Austria and underlying motives and potential barriers regarding ECs, which serve as a basis for designing the questionnaire and the items included in the importance-performance analysis (IPA). Section 3 details the research approach and methodology of the quantitative surveys and IPA. Section 4 presents the findings, and section 5 offers a critical discussion of the results in the context of recent literature.

2.2 Theoretical Background

2.2.1 The legal framework of energy communities in Austria

Over the past three decades, EU energy policies have laid the groundwork for the growing presence of ECs. These strategies view ECs as crucial tools for mobilizing actors within the energy system [21]. The adoption of the European Green Deal in 2019 has marked a significant milestone in this trajectory, further solidifying the EU's commitment to achieving carbon neutrality by 2050 [22]. In the context of promoting the establishment and spread of ECs, the Renewable Energy Directive (REDII [2] and REDIII [1]) and the Electricity Market Directive (EMD) [3] have been central to the spread of ECs by allowing participants to transform their roles from mere passive consumers to active contributors to the sustainable energy transition. In Austria, the concept of energy communities found its footing with the implementation of the Austrian Renewables Expansion Act (Erneuerbaren-Ausbau-Gesetz, EAG) in 2021. Before enforcement of the EAG, the Austrian Electricity Act of 2017 allowed residents of one same multi-apartment building to collectively use PV systems, giving citizens the ability to jointly produce and consume electricity while also being exempt from grid fees [23], which can be seen as a very first form of ECs, but limited to one building. The subsequent implementation of the RED [2] expanded this possibility, thus allowing energy sharing between neighboring buildings. That is to say, the EAG formally enabled the creation of ECs through the definition of the legal and configuration requirements for the establishment and implementation of citizen energy communities (CECs) and renewable energy communities (RECs) [4, 23] in Part 4.1, §16a–§16e. Since then, the framework conditions for the functioning, constellation, organization, outputs, and transparency of ECs have been set. Here, ECs are defined as entities with legal personalities that meet the criteria for associations, cooperative societies, partnerships, small or medium enterprises, or corporations, with at least two members, respectively, shareholders [24]. These entities simultaneously generate, consume, store, and sell self-generated renewable energy while also providing them with the option to offer further energy services without entailing a change in legal status for their participants [25]. A central restriction of RECs is that their primary objective cannot be limited to financial benefits but, instead, should focus on providing either social, environmental, or economic benefits to its members and the REC's area of operation [4].

Nonetheless, financial incentives are significant motivators for joining energy communities [26–28] (see also section 2), and are present in Austrian ECs. To give a monetary incentive to citizens in Austria, there are various legal benefits for ECs: In general, electricity costs include the energy price, system charges, taxes, and surcharges [23]. While the energy price varies by supplier, the other components are fixed and legally defined [29]. These include grid utilization, network loss, metering, and grid provision charges, along with taxes and surcharges like the electricity levy, regional community levy, VAT, and green electricity support payments [23, 29]. EC participants now benefit from reduced costs in these areas, enhancing their motivation to join. It is important to note that ECs, no matter their form, do not provide the full energy supply to their members. As ECs cannot disconnect from the grid, obliging them to use a concessionary grid operator [23], the choice of the existing contract with the primary energy supplier remains valid even after adherence to an EC. Therefore, a second supply contract must be concluded with the EC, meaning that members will receive two bills [30].

As of 2024, the EAG is under revision. The purpose thereof is the acceleration of the transition through the facilitation of the establishment, access to, and management of ECs through the allowance of participation in more than one EC, the introduction of market premiums to promote electricity issuing from hydropower, wind power, photovoltaics, solid biomass and biogas, and investment grants for the construction and expansion of photovoltaic systems, electricity storage systems and wind power plants, among other incentives and flexibility promoting measures [31]. Furthermore, both RED and EMD have been implemented in the Austrian legislature and run parallel with national regulations, and Austria has become one of the leading EU countries regarding the deployment of ECs, taking serious steps in the transposition of EC regulations and operationalization and multiplication of ECs [32].

2.2.2 Understanding member motivations and hurdles

A growing body of literature discusses the motives and challenges associated with participating in ECs and other community energy initiatives, such as energy cooperatives. Literature research suggests four primary reasons for participation in ECs: economic, social, ecological, and technological benefits, which we elaborate on in the following paragraphs. However, the emphasis placed on these main motivations and their prioritizing differs within the reviewed research studies.

ECs and other community initiatives are expected to foster the *expansion of renewables*, thereby supporting the global energy transition and reducing CO_2 emissions [15]. Thus, it comes as no surprise that many ECs share a vision of decarbonization, energy efficiency, and sustainability [33], distinguishing themselves from the "Not In My Backyard (NIMBY)" phenomenon often encountered by larger-scale renewables projects [34, 35]. In that context, many individuals are driven to join ECs out of the desire to contribute to the establishment of a carbon-free society capable of meeting the needs of current and future generations. Through their engagement in ECs, citizens are empowered to actively shape the development of secure and decentralized energy systems and their markets [9] facilitated through collective ownership and participation schemes [36]. Moreover, studies have demonstrated a significant correlation between EC memberships, pro-environmentalist beliefs and intentions, and sustainable energy behavior [37, 38]. Thus, the motive for renewable energy is reinforced by a commitment to environmental protection and conservation [39], with concerns about the environmental impact of traditional energy generation as critical predictors for initiative involvement in ECs [40, 41].

Financial considerations strongly encourage sustainable energy behavior [42]. While ECs are explicitly restricted to being solely profit-oriented, it is unsurprising that financial benefits, meaning the expectation of a return on investment [39], constitute a primary motivator for joining an EC [38, 43]. These financial considerations take various forms, including expected reductions in energy-related expenses (as is the case in Austria through grid fee reductions, see section 2.2.1), financial benefits derived from government incentives for PV systems, increased home and property value, selling self-produced energy, and opportunities to invest in renewable generation plants [26, 33, 44–48]. Consequently, ECs can be categorized as social enterprises due to the potential of these citizen-driven initiatives to improve financial, social, and environmental well-being simultaneously [36]. While there is no consensus on whether monetary or environmental motives hold more significance to members [38], studies have identified a positive correlation between the significance of financial motives and the magnitude of the investment made by members [39]. However, it is noteworthy that some research suggests financial motives may be of lesser importance and less strongly associated with EC memberships [38], indicating that members recognize the non-monetary added value of ECs [40].

The third group of motivations pertains to *social benefits*. The legal obligation to pursue social benefits is also enshrined in legislation: According to Article 2, 16c, the RED mandates that ECs prioritize "environmental, economic, or social community benefits for its shareholders or members, or for the local areas where it operates, rather than financial profits" [2]. However, the precise social impact that ECs may deliver and the methods for measuring their impact remain undefined. A detailed overview of the broad variety of social benefits ECs may bring to people and society was given by Bielig et al. [5], who identified 16 main impacts in four constructs. Out of these, we focused on social capital and community empowerment, as these are the ones most directly affecting citizens. We found two main social impacts that may motivate people to join an EC, and these are often mentioned by ECs looking for new members. This is strengthening social cohesion [40, 46, 49–52] and the region becoming a role model, thereby also improving the region's image [26, 53].

Furthermore, ECs can be embedded in broader social movements and the pursuit of socio-ecological goals [36, 54], which enhances their potential to serve as a powerful vehicle for direct and indirect *energy poverty mitigation* [55–57]. By leveraging renewable energy solutions, adopting inclusive economic models (i.e., tariff reductions), and fostering collaborations with housing associations, social institutions, and the private sector, ECs can achieve substantial energy bill reductions for vulnerable populations while also strengthening social safety nets, thereby enhancing community cohesion and resilience. For instance, a study conducted by Parreño-Rodriguez et al. [55] in Getafe, Madrid, provides empirical evidence that ECs can effectively mitigate energy poverty while empowering communities. By installing collective PV units on public buildings, the project met 77% of the electricity demand for 100 residential units, while the remaining energy covered 60% of the building's energy requirements [55]. Combined with community engagement opportunities such as workshops, seminars, and information sessions, this initiative's long-term benefits and sustainability were amplified [55]. Other EC projects across the EU support these findings.[13] However, the effectiveness of ECs as energy poverty alleviators ultimately hinges on differing critical factors: the clarity of the ECs' structure and goals, the shared values within the community, the understanding of energy poverty itself, and the accessibility of the EC for vulnerable population groups. Another prevalent incentive for joining an EC initiative is the pursuit of *supply security and independence* from suppliers and energy market fluctuations. This includes safeguarding households against price volatility and power disruptions while also enhancing their energy self-reliance and energy autarky [58–60]. Indeed, ECs enable its members to generate, store, monitor, and manage their energy, providing them with a greater sense of autonomy, control, and security [61–64]. This is particularly relevant to people residing in remote areas of Austria, which are more prone to experiencing power outages due to natural disasters. Here, ECs offer avenues for robust power generation and supply. While energy adequacy and reliability are essential motives for joining, it is also often described as a barrier by citizens, fearing that ECs, especially

[13]Examples of EC initiatives focusing on energy poverty alleviation are Coopérnico, Sun4All, BürgerEnergieBerlin, Som Energia, ZEZ, Énergie Solidaire, or POWEpower, among others.

smaller ones, will not be able to provide enough energy for the community in its entirety, leaving its members undersupplied [27, 65, 66]. However, in reality, residual electricity comes from the energy provider, who ensures a continuous supply at all times. Thus, this reluctance may be linked to a lack of information and misunderstandings about how the energy supply within ECs functions. This reliability skepticism is further exacerbated by local demographic specificities, with elderly individuals being more prone to showing less interest in ECs, general skepticism, and disbelief (wait-and-see and NIMBY attitudes), strengthening psychological barriers to joining ECs [65, 66].

While *grid support* is vital to grid operators, thus serving as a central stakeholder motivator [67], it also benefits EC members. For example, grid-friendly EC operations can prevent rising fees linked to necessary network expansions. Numerous studies have explored the optimization of ECs and their energy use, focusing on flexibilities and EC configurations (e.g. [68–75]), highlighting that local energy generation and allocation has the potential to reduce peak demand and alleviate the pressure from the primary grid by decreasing its power flow [76]. Weckesser et al. [68] even suggest a grid loading reduction of up to 58%. However, optimizing an EC with the goal of grid reduction will, in many cases, come with a decrease in prosumers' self-sufficiency rate [70], which could lead to resistance from community members. In that context, most studies primarily focus on technical features, neglecting user preferences and behavior, which could limit the exploitation of the full potential of grid reduction strategies.

Despite the multitude of positive aspects linked to ECs, a few hindering hurdles and concerns persist, hampering the successful establishment and operation of ECs. For instance, transposing supranational EU rules from the RED and EMD into national Austrian law has exposed significant barriers and loopholes.

The intense *administrative effort* required for founding, joining, and coordinating an EC [77–79] represents a significant hurdle. This complex administrative process, which includes extensive paperwork, legal formalities, and coordination with multiple authorities [78], mirrors broader challenges in Austrian public administration, contributing to *structural constraints*, which are further deepened by existing *institutional inertia* [77]. This complexity is not in line with the EU Commission's best practice recommendations for the implementation of ECs, which advocates for accessible and burden-free administrative and regulatory processes [80]. The aim is to avoid discouraging the formation of ECs, particularly among smaller communities that may lack the necessary resources or expertise to navigate and manage such complex administrative frameworks.

Additionally, structural constraints are linked to *resource mobilization and coordination* [81]. In Austria, these difficulties include securing adequate financial resources, and the scarcity of technical expertise [78], barriers that disproportionally affect smaller communities. Concerning internal coordination, the complexity lies in aligning the differing interests of participants, managing resources, and distributing responsibilities, with these coordination efforts becoming increasingly complex as the size of the EC grows [78]. External coordination involves navigating interactions with grid operators, regulatory bodies, and financial institutions, which is further complicated by lacking communication channels [78].

Furthermore, the success of ECs is heavily dependent on *local institutional capabilities*, referring to their readiness to navigate complex regulatory environments, mobilize resources, and coordinate effectively with various stakeholders [78, 82]. These capabilities are frequently constrained by existing regulatory and institutional barriers [77]. For example, national regulations, typically designed for more extensive and more centralized energy systems, do not always translate well to decentralized energy projects at the local scale. The lack of *regulatory clarity* [83] can be attributed to non-binding elements within the legal framework, creating potential loopholes. For instance, technical grid changes by the grid operator, such as infrastructure upgrades or reconfigurations, could disrupt or even terminate existing ECs. While Austrian legislation includes annotations stating that such grid changes should not hinder the operation of established ECs, these annotations are not legally binding and merely serve as guidance, introducing uncertainty for EC developers and participants [78]. This uncertainty is strengthened by an overall *lack of regulatory safeguards and explicit provisions* that protect ECs against financial and operational risks triggered by adverse grid modifications, leaving ECs vulnerable to disruptions and undermining their long-term stability and growth (Fina and Fechner 2021).

In that context, Fina and Fechner [78] emphasize the necessity of involving a diverse range of stakeholders, including project developers, scientists, energy suppliers, and grid operators, to guarantee a successful law-making process that effectively identifies and addresses the loopholes and barriers arising through the transposition of EU regulations. Additionally, Irshaid et al. [77] highlight the critical need for vertical policy integration, which means aligning policies across governance levels to avoid disjointed efforts and enhance the effectiveness of local ECs. Finally, *data privacy and cybersecurity* pose significant concerns and barriers for both EC initiators and (pilot) participants [47]. EC operators have unrestricted access to members' personal information, including detailed load profiles that reveal intrinsic details of households' presence and absence days, device usage, consumption patterns, preferences, lifestyle, and economic status [79]. This data access becomes especially critical as ECs can integrate advanced digital and automated technologies for internal management (e.g., Som Comunitats), energy monitoring (e.g., EnergyID, MARS IoT platform), forecasting (e.g., ERIGRID project), peer-to-peer trading (e.g., Local4Local project), and flexibility services allowing the remote control of energy use and operation of appliances (e.g., PowerShaper), all linked to smart grids and distribution systems [84, 85]. The fact that ECs can be entirely run by private individuals, who may lack knowledge and tools for proper data security measures, further amplifies this vulnerability. Consequently, ECs and similar community energy initiatives can be susceptible to security threats such as data breaches or cyber-attacks on ECs and distribution grids [86, 87]. Implementing cybersecurity monitoring tools and early warning systems, as well as empowering distribution system operators (DSOs) with the ability to disconnect electrical generators independently are critical measures to protect data flows, ensure the uninterrupted operation of ECs, and safeguard essential energy-related services (i.e., heating, cooling, and lighting) from malicious interferences [87]. In response to this vulnerability, a coordination group for cyber-security dubbed "CyberCG" was established to strengthen data infrastructure security and foster trust in those systems and networks [88]. Furthermore, EC initiators have highlighted specific hurdles regarding the quality of the collected energy-related data. Research by Heuninckx et al. [47] found that smart meter installation and data collection and interpretation necessitated more time and effort than initially anticipated. In some cases, the quality of the gathered data was further compromised by issues such as low granularity, incomplete or inaccessible datasets, and format inconsistencies [47].

2.3 Methodology

We conducted an importance-performance analysis of ECs based on nine central characteristics and barriers toward ECs identified via a literature review (see Section 2.2). We used the main motives and hurdles regarding ECs in Austria and formulated neutral statements to enable respondents to rate the (expected) performance and the importance of participation within an EC. The following nine items (translated from German) were selected for the analysis:

1. ECs facilitate the growth of renewable energy and thus support the transition toward sustainable energy systems.
2. ECs are largely self-sufficient in terms of electricity.
3. ECs alleviate the need for further expansion of electricity grids.
4. ECs provide a secure and reliable power supply.
5. ECs provide a role model effect.
6. ECs enhance social cohesion among their members.
7. Personal data (e.g., electricity consumption) are kept secure within an EC.
8. EC members can lower their electricity expenses.
9. The administrative effort (e.g., contracts, accounting) required for EC members is manageable.

The respondents were asked to rate their perceived performance and the importance of their participation in the listed aspects. Importance (I) and Performance (P) of the characteristics were assessed with the following questions: (P): "Below are some characteristics. To what extent do you think an energy community fulfills these characteristics?" and (I): "Below are the same characteristics again. Please now indicate how important these characteristics are to you if you were to participate in an energy community." The ratings for both dimensions were collected using a 5-point Likert scale ranging from "1 = not at all fulfilled" to "5 = absolutely fulfilled" for performance, and "1 = not at all important" to "5 = very important" for importance. Prior to this assessment, a brief and straightforward description of ECs was provided to the respondents, as that concept is not fully known by the public in Austria. An EC was explained with the following simplified explanation, accompanied by an illustration (Figure 2.1): "*In the next questions, we will focus on energy communities. Please read the following explanation carefully: A person has a solar panel on their roof. The solar panel generates electricity from the sun. The person does not use all the electricity himself or herself, so some of it remains unused. The person with the solar panel sells the excess electricity to individuals without solar panels. Together, these individuals form an energy community.*"

Data collection took place between April 28 and May 16, 2022. An online panel provider recruited the respondents. Their service was solely utilized for sampling purposes, while the research team handled the conceptualization, implementation, and analysis of the survey. The sampling strategy aimed

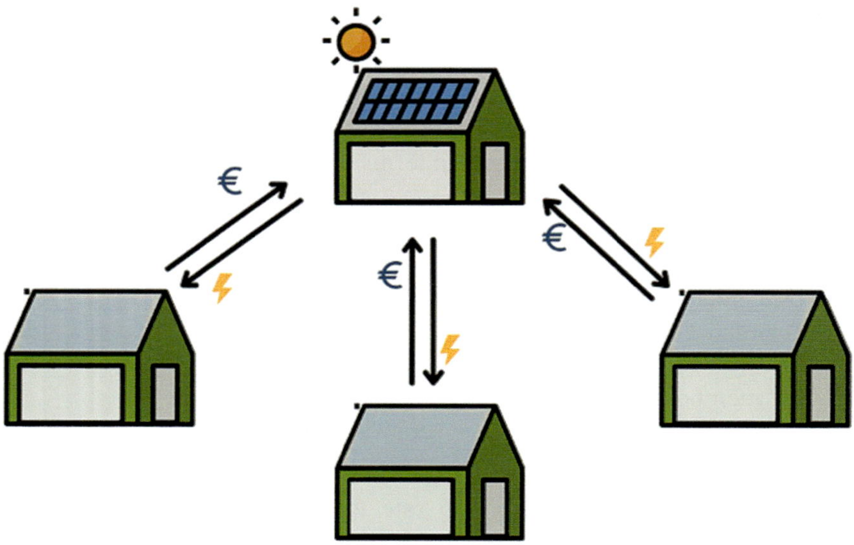

Figure 2.1 Illustration for EC explanation in the survey.

to achieve a representative distribution of respondents across different age groups, genders, and educational backgrounds. However, due to the online-only approach to data collection, the sample cannot be considered fully representative. Despite this, the adequate distribution and diversity in the variables mentioned above offer valuable insights into the perspectives of Austrian residents, which we found adequate to address our research questions. Future research could consider including alternative sampling methods to capture a broader range of perspectives. The questionnaire targeted potential future members of ECs, encompassing the general public without any prerequisite knowledge or existing EC membership. All responses were collected anonymously. A total of 1,073 respondents were included in the analysis. The dataset, including the exact wording of all questions, is provided within the supplementary material. Data was analyzed using IBM SPSS Statistics 29.0. For the importance-performance analysis, the data-centred approach was chosen. Based on the average rating of each dimension (performance and importance), the nine EC characteristics were assigned to the following four categories: "concentrate here" (performance below average, importance above average), "keep up the good work" (performance and importance above average), "low priority" (performance and importance below average), and "possible overkill" (performance above average, importance below average) [89]. In the first step, the means of both performance and importance ratings were calculated, accompanied by paired-sampled t-tests and analyses of variance to assess whether there were significant differences between the performance and importance means.

2.4 Results

2.4.1 Descriptive statistics

Table 2.1 shows the descriptive statistics of the sample. Besides socio-demographic basic information, respondents were also asked questions on energy communities in general. Table 2.2 shows the summary of the results regarding ECs. This gives a good impression of the awareness and perception of ECs in general. The willingness to join an EC is also analyzed deeper in Section 2.4.3.

Regarding Table 2.2, the results show that a majority of 57% of the respondents have never heard about ECs before and have no information about their characteristics. In comparison, 43% have prior knowledge of the subject. Those who have not yet had any points of contact with ECs can only rely on the brief description and depiction (Figure 2.1) given in the questionnaire. Thus, their answers can be regarded as widely unbiased.

The declarations of both respondents, with and without prior knowledge about ECs, to the second question that Table 2.2 refers to, whether they can imagine joining an EC, a majority of 86.5% decided in favor of the answers

Table 2.1 Descriptive statistics of the sample (n = 1,073).

Factor		Percentage
Gender	Male	45.5
	Female	54.4
Age	18–29	18.4
	30–39	19.5
	40–49	20.8
	50–59	23.9
	60–69	17.5
Education	Without Matura*	49.9
	With Matura*	50.1
Income	Quartile Q1	39.0
	Quartile Q2	24.6
	Quartile Q3	22.5
	Quartile Q4	13.9

*Secondary school leaving certificate in AT, corresponding to A-levels

Table 2.2 Knowledge and willingness to join an EC (n = 1,073).

Question	Response	Percentage
Have you already heard of energy communities?	Yes	43.1
	No	56.9
Could you generally imagine joining an energy community in your neighborhood?	Yes	46.0
	Maybe	40.5
	No	8.5
	I do not know	4.9

"yes" and "maybe". Therefore, 86.5% of the respondents are generally open to participating in an EC. Furthermore, 8.5% of the respondents cannot imagine joining an EC, while 4.9% replied with "I don't know." With this differentiation, it was possible to explore further the effect of the variable "willingness to join an EC" on the perception of performance and importance values discussed in Section 2.4.3.

2.4.2 Importance of performance analysis

The results of the importance-performance analysis are depicted in Table 2.3, with the arithmetic mean and standard deviations (SD) of both dimensions, performance and importance, as well as t-test results. The mean ratings for both dimensions are above the scale average, which implies that all characteristics are expected to be fulfilled by an EC and are essential for an individual's EC membership.

The highest-rated aspect, receiving an average score of 4.03, pertains to the support for renewable energy expansion, which respondents anticipate will be facilitated by implementing energy communities (ECs). Financial gain follows closely with an average rating of 3.83, with approximately 70% of respondents expressing belief in potential cost savings through EC membership. Conversely, the lowest-rated aspect is the security of personal data, receiving an average score of 3.44. Half of the respondents either express uncertainty or disagree regarding the safety of their data within ECs.

In terms of the importance of individual participation in ECs, financial gain ranks highest, with a score of 4.43. Only 2% of respondents consider savings on energy bills to be insignificant. Following closely are supply security (4.40) and achieving autonomy in power supply (4.19). Support for expanding renewable energy and facilitating the energy transition ranks fourth with an average score of 4.11. Social aspects such as serving as a role model (3.61) and promoting social cohesion (3.65) are considered less important in joining an EC.

Table 2.3 Results of performance and importance rating.

Factor	Performance		Importance		Mean difference	t-value	p-value
	mean	SD	mean	SD			
Support of renewables	4.03	0.82	4.11	0.89	–0.08	–3.197 **	<0.001
Autonomy	3.73	0.89	4.19	0.83	–0.46	–15.864 **	<0.001
Grid support	3.63	0.96	3.87	0.95	–0.23	–7.838 **	<0.001
Supply security	3.49	0.90	4.40	0.81	–0.91	–28.621 **	<0.001
Role model	3.79	0.89	3.61	1.04	0.17	5.756 **	<0.001
Social cohesion	3.70	0.84	3.65	1.01	0.05	1.828	0.068
Financial gain	3.83	0.84	4.43	0.77	–0.59	–21.393 **	<0.001
Administrative burden	3.51	0.86	4.05	0.81	–0.54	–16.801 **	<0.001
Data security	3.44	0.88	4.08	0.97	–0.64	–18.910 **	<0.001

T-test two-tail with probability $*p < 0.05$ or $**p < 0.01$

Mean scores measured on a 5-point Likert scale (from 1 = low to 5 = high importance/performance)

Mean difference = satisfaction mean – importance mean

As described in the methodology and depicted in Figure 2.2, the nine items were categorized into four quadrants based on their mean ratings. It is crucial to emphasize that these results solely reflect the perceived performance of these aspects among potential EC members and may not align with the actual performance of existing ECs. Perceptions among respondents may be influenced by misinformation and a lack of comprehensive understanding of ECs, particularly pronounced due to an information deficit in Austria. This gap in public discourse and information dissemination can lead to either inflated expectations or underestimations of EC effectiveness, with corresponding implications discussed in the following section.

2.4.3 Importance performance analysis based on willingness to join an energy community

The results above represent the whole sample of 1,073 respondents. In addition, we aimed to explore how interest in joining ECs influences the participants' perception of EC attributes and to identify specific characteristics or barriers that are particularly significant obstacles for those who are hesitant to participate. For this aim, the importance-performance analysis was repeated for the variable "willingness to join an EC," for which the

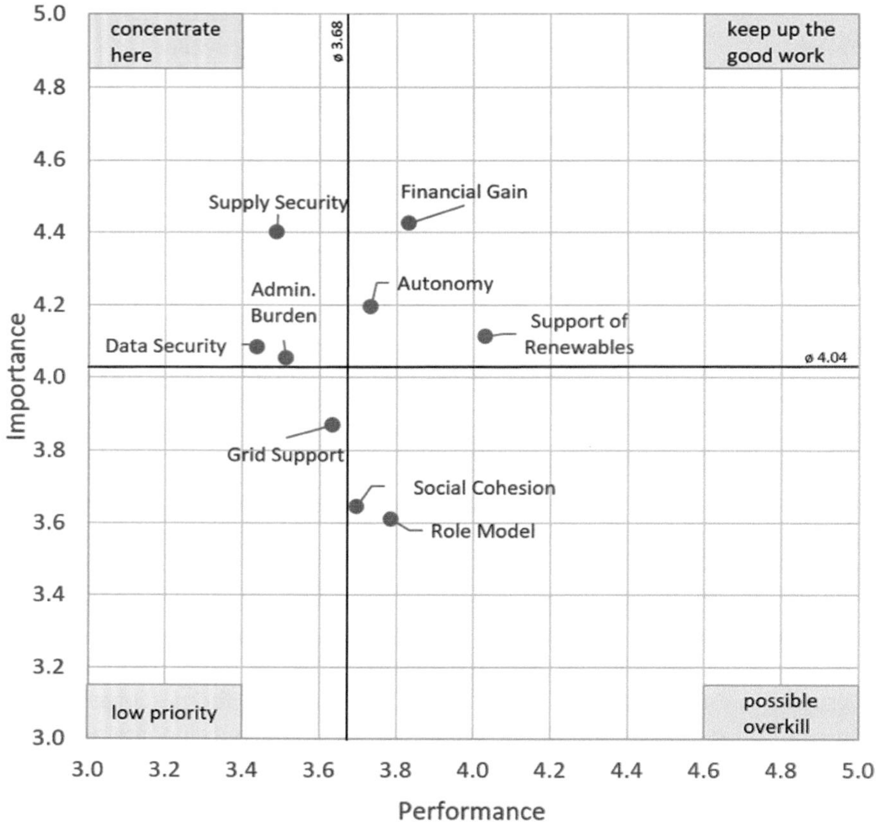

Figure 2.2 Importance-performance analysis on EC characteristics (n = 1,073)

ANOVA shows significant differences in the perception of performance and importance values for all nine items. Respondents are separated into three groups based on their willingness to join an EC, namely high willingness (respondents who answered "yes" to the respective question, n = 494), middle willingness (respondents who answered "maybe", n = 435) and low willingness (respondents who answered "no," n = 91). The results of descriptive statistics and ANOVA are shown in Table 2.4.

Figure 2.3, Figure 2.4, and Figure 2.5 depict the results of the IPA for each of the three interest groups separately. While all characteristics show significant group differences, those characteristics that change the assignment to the IPA quadrants are of particular interest (see Table 2.5). It becomes

Table 2.4 Results of ANOVA based on willingness to join an EC.

	Financial satisfaction	Performance				Importance			
		Mean	SD	F	p-value	Mean	SD	F	p-value
Support of renewables	Yes	4.29	0.709	66.150**	<0.001	4.38	0.75	61.524**	<0.001
	Maybe	3.91	0.753			4.01	0.85		
	No	3.37	1.082			3.38	1.21		
Autonomy	Yes	3.62	0.854	28.375**	<0.001	4.13	0.78	34.671**	<0.001
	Maybe	3.56	0.845			4.06	0.91		
	No	3.27	1.056			3.64	1.20		
Grid support	Yes	3.64	0.955	25.011**	<0.001	4.55	0.70	22.142**	<0.001
	Maybe	3.42	0.880			3.85	0.82		
	No	3.08	1.079			3.30	1.13		
Supply security	Yes	3.64	0.878	19.585**	<0.001	4.55	0.70	28.079**	<0.001
	Maybe	3.42	0.841			3.89	1.16		
	No	3.04	1.084			3.09	1.16		
Role model	Yes	3.67	0.835	70.025**	<0.001	3.53	0.99	23.151**	<0.001
	Maybe	3.67	0.865			3.75	1.02		
	No	3.02	1.011			3.04	1.19		
Social cohesion	Yes	3.67	0.765	38.410**	<0.001	3.75	1.02	12.941**	<0.001
	Maybe	3.67	0.765			3.62	0.93		
	No	3.05	1.089			3.18	1.23		
Financial gain	Yes	4.08	0.737	56.620**	<0.001	4.54	0.67	22.057**	<0.001
	Maybe	3.73	0.900			4.40	0.72		
	No	3.19	1.064			3.99	0.45		
Administrative burden	yes	3.49	0.807	27.598**	<0.001	4.10	0.74	13.411**	<0.001
	Maybe	3.49	0.807			3.65	1.02		
	No	2.95	0.856			3.30	1.12		
Data security	Yes	3.57	0.851	19.616**	<0.001	4.12	0.99	5.288**	0.005
	Maybe	3.39	0.975			4.10	0.90		
	No	2.97	0.960			3.77	1.18		

ANOVA with probability $^*p < 0.05$ or $^{**}p < 0.01$

evident that for people with a medium interest in joining, which is 40.5% of the sample, more characteristics are of concern than people with a high willingness to join. While supply security is unsatisfactory for all groups, for those with medium interest in joining, especially autonomy, financial gain, as well as administrative burden, and data security are key barriers that need to be addressed. For all those elements, perceived performance is below the importance ratings, resulting in an unsatisfactory service offering from ECs.

The prioritization matrix of the IPA further highlights a notable disconnect between technical necessities, such as grid support, and perceived priorities within ECs. While supply security is classified as a "concentrate here" element, indicating it is highly important while currently performing, grid support is rated as "low priority" or even "possible overkill" despite its critical functions in preventing outages and ensuring seamless integration

Table 2.5 Overview of IPA results based on willingness to join an EC

	High willingness (answer = yes)	**Medium willingness (answer = maybe)**	**Low willingness (answer = no)**
Support of renewables	Good work	Good work/overkill	Good work/overkill
Autonomy	Good work	**Concentrate**	Good work
Grid support	Low priority	Low priority	Low priority/overkill
Supply security	**Concentrate**	**Concentrate**	**Concentrate**
Role model	Low priority	Low priority	Overkill
Social cohesion	Low priority	Low priority	Overkill
Financial gain	Good work	**Concentrate**	Good work
Administrative burden	Low priority	**Concentrate**	**Concentrate**
Data security	Low priority	**Concentrate**	**Concentrate**

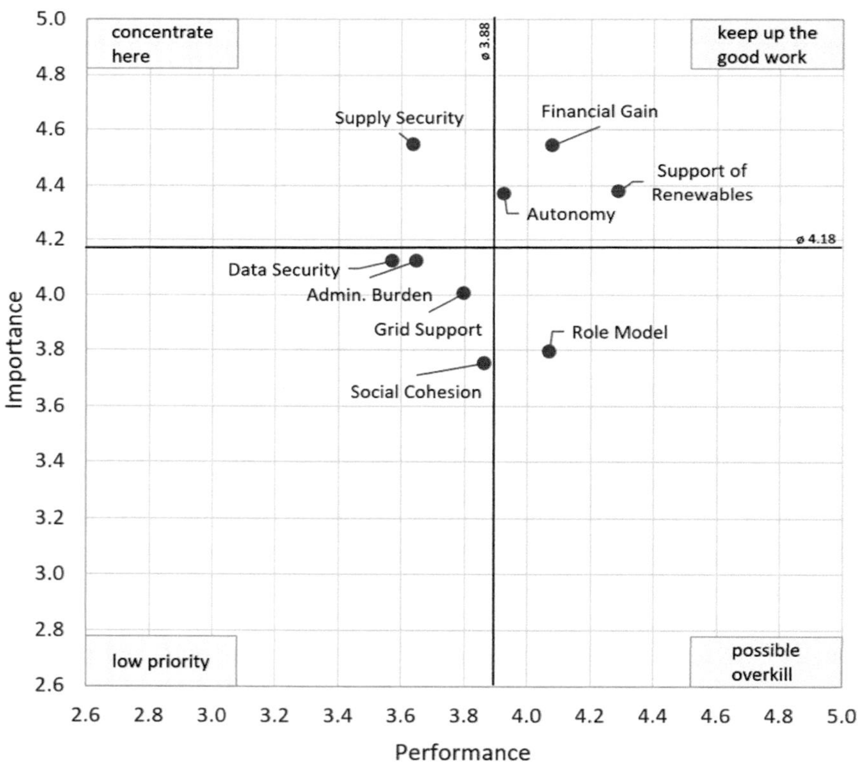

Figure 2.3 Results of IPA for high willingness to join an EC (n = 429).

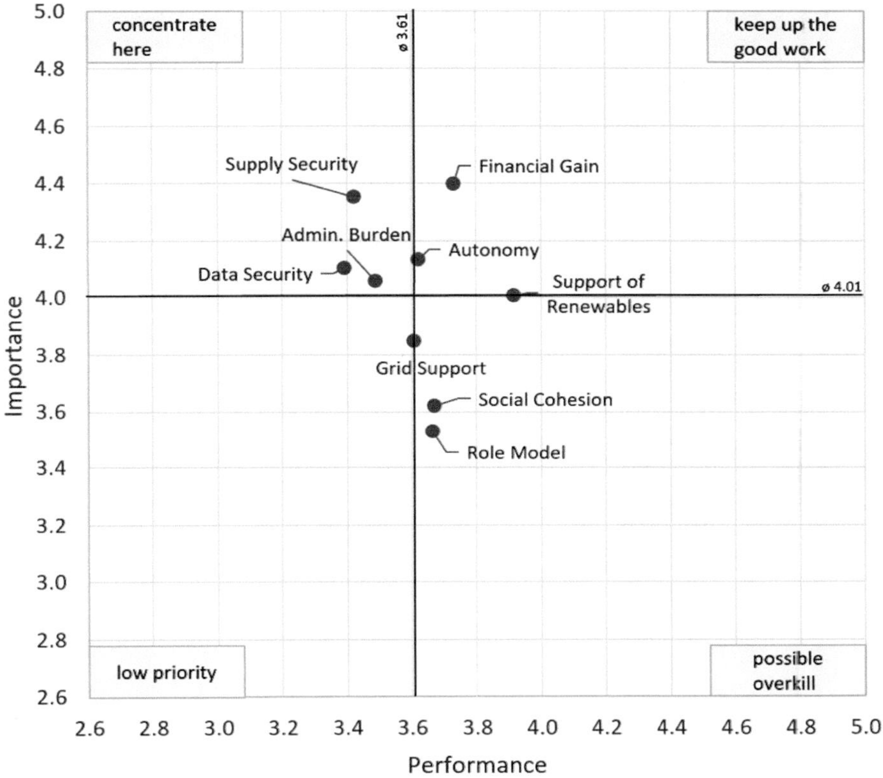

Figure 2.4 Results of IPA for medium willingness to join an EC (n = 435).

of renewable energy sources. This misalignment suggests that key technical components critical for the functioning and long-term reliability of ECs are potentially being undervalued, likely due to limited immediate visibility to end-users.

Moreover, support of renewables, which is perceived as "good work" by the first group, diminishes between "good work" and "possible overkill," which means that this argument is not as important for medium- and low-interested people. For the social components, namely role model and social cohesion, importance and performance decreases. Still, they are assigned to all three groups in either "low priority" or "possible overkill," which undermines their minor role. Not only is there a shift of characteristics to concentrate on for those with low willingness to join. In addition, the average

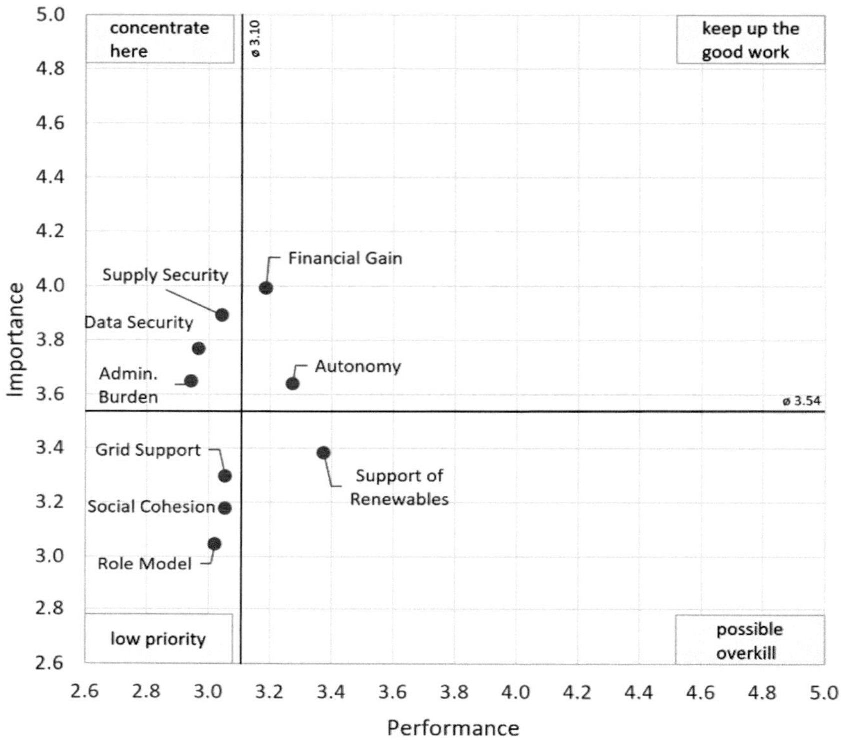

Figure 2.5 Results of IPA for high willingness to join an EC (n = 91).

agreement to all statements is significantly lower. That is to say, with a lower intent to join an EC, the perception of EC performance is declining – or vice versa.

2.5 Discussion and Conclusion

Our data, generated in 2022, reveals a surprisingly high level of interest in and an overall optimistic viewpoint on ECs, despite them being a new possibility and information about ECs being relatively low to this time. Notably, only a small share of respondents (8.9%) stated that they were not interested in joining an EC, while a majority showed great openness to joining (86.5% would join or maybe join). To capitalize on realizing this potential and sustaining interest, it is imperative to provide accurate information and engage in robust educational efforts to not only maintain interest in community energy projects but also ensure that this majority is not lost during the implementation phase of ECs in Austria.

One important distinction to common IPA studies conducted in marketing research is that within this study, the expected performance of ECs is evaluated rather than the actual performance based on experience. This approach assesses preconceptions of a service rather than actual experiences. Even more, the "service" offered by ECs is still innovative and uncommon in Austria. This also limits our study, as it is unclear whether respondents truly understood the concept of an EC. Indeed, nearly 57% of the respondents stated that they had never heard of ECs before being involved in this study (Table 2.2). On the other hand, this is also one strength of our study: much research toward EC is conducted among members of already existing EC initiatives. This often involves particularly interested individuals whose views are not necessarily representative of the entire population.

Regarding the investigated motives and barriers, we organize this discussion along the four quadrants of the IPA, starting with the section "keep up the good work". This section illustrates characteristics considered essential for membership in energy communities (ECs), which are already perceived to be fulfilled. These findings align with existing literature [33, 38, 43, 62–64]. However, gaps in knowledge persist regarding *autonomy* and *financial gains*, potentially leading to unrealistic expectations and prolonged dissatisfaction, as ECs typically offer only partial energy autonomy and modest cost savings. Further research is needed to determine the expected and acceptable levels of savings among participants. While ECs are generally perceived as contributing positively to the *energy transition*, there is a lack of comprehensive long-term evaluations on the actual environmental impact of ECs. Photovoltaics, in particular, could play an important role for energy communities in the future though and represent a major opportunity for the EU's green energy transition, which promotes corresponding incentive systems and ascribes a significant share of decentralized energy generation to the technology [90]. In 2024, the Austrian Coordination Centre for Energy Communities registered a 75% share of energy communities in Austria providing photovoltaic electricity, while the remaining 25% consists of photovoltaic combinations with wind power, hydropower, and biomass [91]. Consequently, it remains unclear whether ECs promote the establishment of new PV systems or primarily reframe energy distribution by utilizing existing generation plants within ECs.

The most significant focus should be directed toward the section "concentrate here", which features items of considerable importance but are perceived to perform relatively poorly. This is particularly noteworthy as deficiencies in *supply security*, *data security*, and *administrative burden* are perceived as significant barriers to attracting new members and dampening citizens' willingness to participate [47, 77–79]. While these aspects are crucial for fostering participation, our results show a prevailing concern among individuals that their expectations may not be met. Thus, enhancing the performance of these attributes is crucial to mitigate the risk of deterring potential members. Supply security and data security are critical aspects that demand significant attention. Despite their importance as motivators for participation [40, 47, 58, 64, 79], their perceived performance falls below expectations in our study. For example, participants need an improved understanding regarding residual current supply from utilities to clarify that supply security remains comparable to traditional central power supply systems. This barrier can be addressed through targeted awareness campaigns on supply security within ECs. Data security is equally vital for EC participants, given the pervasive digitalization involved. Concerns may arise due to the potential involvement of private individuals in EC management who may lack adequate knowledge and tools to meet stringent data security standards. Moreover, apprehension toward (larger) companies like grid operators, who have access to smart meter data within ECs, can further exacerbate mistrust. Further research is essential to explore these concerns thoroughly and develop solutions to mitigate existing data security challenges. Another significant factor to consider is the administrative burden perceived as high by respondents. Minimizing this burden is crucial to attracting new members. This concern is also highlighted by Neumann et al. [92], who recommend that EC service providers streamline the participation process, given the current complexity of contractual agreements that may deter individuals from joining an EC.

In the "low priority" category, only one item is identified: *grid support*. Interestingly, while supply security is categorized as "concentrate here," grid support, despite being the backbone of our contemporary energy systems [93], receives a comparatively lower prioritization. While this outcome was anticipated, as it does not significantly influence people's willingness to participate [70], it is still surprising given that grid support is critical for maintaining energy stability and directly bolstering supply security, playing a key role in advancing the clean energy transition by enabling the integration and large-scale deployment of renewable energy sources [93]. This suggests a potential lack of understanding of how these technical aspects are intertwined to ensure the reliability of EC operations. While the "low priority" rating for grid support might imply that communication efforts on this aspect should receive minimal emphasis, it is noteworthy that individuals still prioritize ECs being grid-friendly over aspects like promoting social cohesion, which is classified as "possible overkill." Bridging this perception gap between grid support and supply security requires targeted educational initiatives, such as specialized workshops, informational campaigns, and ongoing training for both EC stakeholders and end-users. By fostering a deeper understanding of the technical foundations of energy stability and reliability, notably less visible aspects such as grid support, ECs can establish a framework that aligns technical needs with the public perception of EC functioning.

The formerly mentioned section ("possible overkill") gave way to interesting results. The two aspects regarding social benefits, namely *role model* and *social cohesion*, which are often discussed in EC narratives and demanded by the RED, are of the least importance for prospective members. This result contradicts literature findings in [26, 46, 59, 94], and these aspects are already perceived as fulfilled. As a result, these factors currently do not serve as decisive criteria for joining an EC, and increased information dissemination on these subjects may meet the lack of interest among potential customers. Notably, while social aspects are still deemed somewhat important within this study, they rank lower compared to all other characteristics, including grid support, in terms of respondents' interest in participating in an EC. On the other hand, to foster the social impact and value of ECs, other social aspects, such as the support of vulnerable groups, may be decisive for joining an EC. However, evidence on the actual social impact of ECs is currently lacking [5], and although the Austrian EAG and RED claim for social community benefits (Art 2, 16b), concise requirements on the social impact are not available.

Further, our study investigated barriers and motives across three distinct subgroups categorized by their interest ECs, shedding light on diverse perceptions and priorities that may impact participation decisions. While the support of renewables is viewed favorably among respondents with high interest, its perceived importance diminishes progressively across respondents with a lower willingness to join. Similarly, social aspects (role model and social cohesion) exhibit decreasing importance and performance ratings across all groups, positioning them in "low priority" or "possible overkill" categories and highlighting their minimal impact on participation decisions. Moreover, individuals with lower interest in joining an EC not only prioritize different characteristics but also demonstrate significantly lower average agreement with all statements compared to their more interested counterparts. This indicates a direct correlation between intent to join and perception of

EC performance (or vice versa), suggesting that improving these perceptions could potentially enhance overall willingness to participate.

These findings emphasize the importance of targeted strategies tailored to address diverse concerns among potential EC members. While core motivations such as financial gain and energy autonomy resonate strongly across the sample, understanding and mitigating barriers related to administrative complexity and data security are paramount. Furthermore, effectively communicating the environmental benefits of ECs while recognizing varying perceptions across different interest groups is crucial for fostering broader acceptance and participation in energy community initiatives. Successful education and sensitization efforts, particularly regarding the benefits and realistic expectations of ECs, are vital to maintaining a high level of interest and ensuring the long-term sustainability of EC initiatives. One example is the EU-funded SHAREs initiative that supports individuals interested in founding RECs and CECs by providing comprehensive resources through country-specific platforms in Austria, Bulgaria, Croatia, Georgia, Germany, and Hungary [95]. These so-called "local heroes" are equipped with the necessary legal, financial, and technical knowledge through information packages and tailored communication campaigns [95]. By leveraging tools, best practices, and results from existing initiatives, SHAREs prevents new EC initiators from having to start from scratch, thus effectively broadening the engagement of EC projects [95], reaching a wider audience, such as the 57% of our study who were previously unfamiliar with ECs. Community and Renewable Energy Scheme (CARES), funded by the Scottish Government, is designed to support existing and new community-centred renewable energy projects [96]. CARES provides a range of resources, including educational material, workshops and webinars, step-by-step toolkits, and free personalized advice aimed at enhancing the sustainability and effectiveness of community energy initiatives, thereby playing a crucial role in advancing Scotland in reaching its net-zero targets [96]. Energie Partagée is a French education and support initiative dedicated to empowering individuals and communities to actively partake in the renewable energy transition by developing and managing citizen-driven energy projects by offering educational resources and support [97].

Discussing the results of our study, we come to the following significant conclusions and recommendations to streamline citizens' expectations with EC reality:

- Enhance transparency and education on EC benefits and limitations, energy autonomy, and financial savings to set realistic participation expectations and prevent potential disappointment.

- Prioritize efforts to improve supply security and data protection within ECs through targeted awareness-raising and educational measures.
- Streamline administrative processes and reduce bureaucratic burdens to increase accessibility to EC initiation and membership, as highlighted by current participant concerns.
- Close existing loopholes and provide regulatory safeguards to protect ECs against financial and operational risks.
- Conduct comprehensive, long-term evaluations to assess the true environmental impact of ECs.
- Focus research efforts on understanding and enhancing the social impacts of ECs, including support for structurally excluded and vulnerable groups, to align with regulatory expectations and community needs.

To fully leverage these findings and foster continued growth and engagement in ECs in Austria, it is crucial to align citizens' expectations with the practical realities of participation while addressing existing barriers and communicating the full scope of benefits through targeted communication and sensitization campaigns. Further research is needed to explore the long-term environmental impacts of ECs, the role of social inclusivity and supporting structurally excluded and vulnerable groups as a lever to increase participation, and ways to simplify administrative processes in ECs to encourage broader participation and long-term engagement.

Acknowledgments

Author contribution statement: Frederike Ettwein: Conceptualization, Methodology, Investigation, Formal analysis, Writing – Original Draft, Writing – Review and Editing; Mélanie Michel: Writing – Original Draft, Writing – Review and Editing, Flora Bachleitner: Writing – Original Draft, Writing – Review and Editing.

References

[1] Directive (EU) 2023/2413. Directive (EU) 2023/2413 of the European Parliament and of the Council of 18 October 2023 amending Directive (EU) 2018/2001, Regulation (EU) 2018/1999 and Directive 98/70/EC as regards the promotion of energy from renewable sources, and repealing Council Directive (EU) 2015/652 2023.

[2] Directive (EU) 2018/2001. Directive (EU) 2018/2001 of the European Parliament and of the Council of 11 December 2018 on the promotion of the use of energy from renewable sources (recast) (Text with EEA relevance.) 2018.

[3] Directive (EU) 2019/944. Directive (EU) 2019/944 of the European Parliament and of the Council of 5 June 2019 on common rules for the internal market for electricity and amending Directive 2012/27/EU (recast) (Text with EEA relevance.) 2019.

[4] EAG 2021. Bundesgesetz über den Ausbau von Energie aus erneuerbaren Quellen (Erneuerbaren-Ausbau-Gesetz—EAG). F.L.G. I 150/2021. 2021.

[5] Bielig M, Kacperski C, Kutzner F, Klingert S. Evidence behind the narrative: Critically reviewing the social impact of energy communities in Europe. Energy Research & Social Science 2022;94:102859. https://doi.org/10.1016/j.erss.2022.102859.

[6] Campos I, Marín-González E. People in transitions: Energy citizenship, prosumerism and social movements in Europe. Energy Research & Social Science 2020;69:101718. https://doi.org/10.1016/j.erss.2020.101718.

[7] Chaudhry S, Surmann A, Kühnbach M, Pierie F. Renewable Energy Communities as Modes of Collective Prosumership: A Multi-Disciplinary Assessment, Part I—Methodology. Energies 2022;15:8902. https://doi.org/10.3390/en15238902.

[8] Broska LH, Vögele S, Shamon H, Wittenberg I. On the Future(s) of Energy Communities in the German Energy Transition: A Derivation of Transformation Pathways. Sustainability 2022;14:3169. https://doi.org/10.3390/su14063169.

[9] Cappellaro F, D'Agosta G, De Sabbata P, Barroco F, Carani C, Borghetti A, et al. Implementing energy transition and SDGs targets throughout energy community schemes. Journal of Urban Ecology 2022;8:juac023. https://doi.org/10.1093/jue/juac023.

[10] European Environment Agency. Energy prosumers in Europe: citizen participation in the energy transition. LU: Publications Office; 2022.

[11] Gjorgievski VZ, Cundeva S, Georghiou GE. Social arrangements, technical designs and impacts of energy communities: A review. Renewable Energy 2021;169:1138–56. https://doi.org/10.1016/j.renene.2021.01.078.

[12] Eichman J, Torrecillas Castelló M, Corchero C. Reviewing and Exploring the Qualitative Impacts That Different Market and Regulatory Measures Can Have on Encouraging Energy Communities Based on

Their Organizational Structure. Energies 2022;15:2016. https://doi.org/10.3390/en15062016.
[13] Van Veelen B. Negotiating energy democracy in practice: governance processes in community energy projects. Environmental Politics 2018;27:644–65. https://doi.org/10.1080/09644016.2018.1427824.
[14] Mutani G, Santantonio S, Brunetta G, Caldarice O, Demichela M. An energy community for territorial resilience: Measurement of the risk of an energy supply blackout. Energy and Buildings 2021;240:110906. https://doi.org/10.1016/j.enbuild.2021.110906.
[15] Ahmed S, Ali A, D'Angola A. A Review of Renewable Energy Communities: Concepts, Scope, Progress, Challenges, and Recommendations. Sustainability 2024;16:1749. https://doi.org/10.3390/su16051749.
[16] Neves C, Oliveira T, Sarker S. Citizens' participation in local energy communities: the role of technology as a stimulus. European Journal of Information Systems 2024.
[17] Nagpal H, Avramidis I-I, Capitanescu F, Madureira AG. Local Energy Communities in Service of Sustainability and Grid Flexibility Provision: Hierarchical Management of Shared Energy Storage. IEEE Trans Sustain Energy 2022;13:1523–35. https://doi.org/10.1109/TSTE.2022.3157193.
[18] Liu Y, Wang Y, Xi H, Lin J, Ma J. Community Energy Cooperation with Shared Energy Storage for Economic-Environment Benefits. 2022 IEEE PES Innovative Smart Grid Technologies - Asia (ISGT Asia), Singapore, Singapore: IEEE; 2022, p. 230–4. https://doi.org/10.1109/ISGTAsia54193.2022.10003636.
[19] EbUtilities. Marktpartner-Verzeichnis. Online Request on registered RECs and CECs in Austria 2024. https://www.ebutilities.at/marktpartner (accessed April 17, 2024).
[20] Hampl N, Sposato R, Marterbauer G, Nowshad A, Strebl M, Salmhofer A, et al. Erneuerbare Energien in Österreich 2022 - Der jährliche Stimmungsbarometer der österreichischen Bevölkerung zu erneuerbaren Energien. Institut für Produktions-, Energie- und Umweltmanagement, Alpen-Adria-Universität Klagenfurt; 2022.
[21] Houtman A, Reins L. Energy Transition in the EU: Targets, Market Regulation and Law. In: Wood G, Onyango V, Yenneti K, Liakopoulou MA, editors. The Palgrave Handbook of Zero Carbon Energy Systems and Energy Transitions, Cham: Springer International Publishing; 2022, p. 1–26. https://doi.org/10.1007/978-3-030-74380-2_2-2.
[22] Widuto A. Energy transition in the EU. Brussels: | European Parliamentary Research Service; 2023.

[23] ElWOG 2010. Bundesgesetz, mit dem die Organisation auf dem Gebiet der Elektrizitätswirtschaft neu Geregelt Wird (Elektrizitätswirtschafts- und –Organisationsgesetz 2010—ElWOG 2010). F.L.G. I No. 110/2010 last amended by F.L.G. I No. 150/2021.. 2010.
[24] Moroni S, Alberti V, Antoniucci V, Bisello A. Energy Communities in a Distributed-Energy Scenario: Four Different Kinds of Community Arrangements. In: Bisello A, Vettorato D, Laconte P, Costa S, editors. Smart and Sustainable Planning for Cities and Regions, Cham: Springer International Publishing; 2018, p. 429–37. https://doi.org/10.1007/978-3-319-75774-2_29.
[25] Biresselioglu ME, Limoncuoglu SA, Demir MH, Reichl J, Burgstaller K, Sciullo A, et al. Legal Provisions and Market Conditions for Energy Communities in Austria, Germany, Greece, Italy, Spain, and Turkey: A Comparative Assessment. Sustainability 2021;13:11212. https://doi.org/10.3390/su132011212.
[26] Conradie PD, Ruyck OD, Saldien J, Ponnet K. Who wants to join a renewable energy community in Flanders? Applying an extended model of Theory of Planned Behaviour to understand intent to participate. Energy Policy 2021;151:112121. https://doi.org/10.1016/j.enpol.2020.112121.
[27] Hicks J, Ison N. An exploration of the boundaries of 'community' in community renewable energy projects: Navigating between motivations and context. Energy Policy 2018;113:523–34. https://doi.org/10.1016/j.enpol.2017.10.031.
[28] Soeiro S, Ferreira Dias M. Community renewable energy: Benefits and drivers. Energy Reports 2020;6:134–40. https://doi.org/10.1016/j.egyr.2020.11.087.
[29] SNE-V 2018. Verordnung der Regulierungskommission der E-Control, mit der die Entgelte für die Systemnutzung bestimmt werden (Systemnutzungsentgelte-Verordnung 2018 – SNE-V 2018) 2018.
[30] Österreichische Koordinationsstelle für Energiegemeinschaften. Erneuerbare Energiegemeinschaften für Haushalte. Klima und Energiefonds; 2024.
[31] Austrian Parliament. Ministerialentwurf: Elektrizitätswirtschaftsgesetz, Energiearmuts-Definitions-Gesetz; Energie-Control-Gesetz, Änderung (310/ME). Austrian Parliament 2024. https://www.parlament.gv.at/gegenstand/XXVII/ME/310?selectedStage=100 (accessed April 23, 2024).
[32] REScoop. Austria - Enabling Frameworks & Support Schemes. REScoop 2022. https://www.rescoop.eu/policy/transposition-tracke

r/enabling-frameworks-support-schemes/austria (accessed June 6, 2024).

[33] Van Der Schoor T, Scholtens B. Power to the people: Local community initiatives and the transition to sustainable energy. Renewable and Sustainable Energy Reviews 2015;43:666–75. https://doi.org/10.1016/j.rser.2014.10.089.

[34] O'Neil SG. Community obstacles to large scale solar: NIMBY and renewables. J Environ Stud Sci 2021;11:85–92. https://doi.org/10.1007/s13412-020-00644-3.

[35] Petrova MA. From NIMBY to acceptance: Toward a novel framework — VESPA — For organizing and interpreting community concerns. Renewable Energy 2016;86:1280–94. https://doi.org/10.1016/j.renene.2015.09.047.

[36] Becker S, Kunze C, Vancea M. Community energy and social entrepreneurship: Addressing purpose, organisation and embeddedness of renewable energy projects. Journal of Cleaner Production 2017;147:25–36. https://doi.org/10.1016/j.jclepro.2017.01.048.

[37] Sloot D, Jans L, Steg L. Can community energy initiatives motivate sustainable energy behaviours? The role of initiative involvement and personal pro-environmental motivation. Journal of Environmental Psychology 2018;57:99–106. https://doi.org/10.1016/j.jenvp.2018.06.007.

[38] Sloot D, Jans L, Steg L. In it for the money, the environment, or the community? Motives for being involved in community energy initiatives. Global Environmental Change 2019;57:101936. https://doi.org/10.1016/j.gloenvcha.2019.101936.

[39] Holstenkamp L, Kahla F. What are community energy companies trying to accomplish? An empirical investigation of investment motives in the German case. Energy Policy 2016;97:112–22. https://doi.org/10.1016/j.enpol.2016.07.010.

[40] Soeiro S, Ferreira Dias M. Renewable energy community and the European energy market: main motivations. Heliyon 2020;6:e04511. https://doi.org/10.1016/j.heliyon.2020.e04511.

[41] Goedkoop F, Sloot D, Jans L, Dijkstra J, Flache A, Steg L. The Role of Community in Understanding Involvement in Community Energy Initiatives. Front Psychol 2022;12. https://doi.org/10.3389/fpsyg.2021.775752.

[42] Reisch LA, Thøgersen J, editors. Handbook of research on sustainable consumption. Cheltenham, UK Northampton, MA, USA: Edward Elgar Publishing; 2015.

[43] Doci G, Vasileiadou E, Petersen A. Exploring the transition potential of renewable energy communities. Working Papers 2014.
[44] Doci G. Collective Action with Altruists: How Are Citizens Led Renewable Energy Communities Developed? Sustainability 2021;13:507. https://doi.org/10.3390/su13020507.
[45] Hackbarth A, Löbbe S. What motivates private households to participate in energy communities? A literature review and German case study. Energy Communities 2022:153–66. https://doi.org/10.1016/B978-0-323-91135-1.00026-2.
[46] Schall DL. More than money? An empirical investigation of socio-psychological drivers of financial citizen participation in the German energy transition. Cogent Economics & Finance 2020;8:1777813. https://doi.org/10.1080/23322039.2020.1777813.
[47] Heuninckx S, Meitern M, te Boveldt G, Coosemans T. Practical problems before privacy concerns: How European energy community initiatives struggle with data collection. Energy Research & Social Science 2023;98:103040. https://doi.org/10.1016/j.erss.2023.103040.
[48] Abada I, Ehrenmann A, Lambin X. On the Viability of Energy Communities. The Energy Journal 2020;41:113–50. https://doi.org/10.5547/01956574.41.1.iaba.
[49] Boon FP, Dieperink C. Local civil society based renewable energy organisations in the Netherlands: Exploring the factors that stimulate their emergence and development. Energy Policy 2014;69:297–307. https://doi.org/10.1016/j.enpol.2014.01.046.
[50] Aldrich DP, Meyer MA. Social capital and community resilience. American Behavioral Scientist 2015;59:254–69.
[51] Lin N. Building a network theory of social capital. Social Capital 2017:3–28.
[52] Coy D, Malekpour S, Saeri AK, Dargaville R. Rethinking community empowerment in the energy transformation: A critical review of the definitions, drivers and outcomes. Energy Research & Social Science 2021;72:101871.
[53] Martens K. Investigating subnational success conditions to foster renewable energy community co-operatives. Energy Policy 2022;162:112796. https://doi.org/10.1016/j.enpol.2022.112796.
[54] Kunze C, Becker S. Collective ownership in renewable energy and opportunities for sustainable degrowth. Sustain Sci 2015;10:425–37. https://doi.org/10.1007/s11625-015-0301-0.

[55] Parreño-Rodriguez A, Ramallo-González AP, Chinchilla-Sánchez M, Molina-García A. Community energy solutions for addressing energy poverty: A local case study in Spain. Energy and Buildings 2023;296:113418. https://doi.org/10.1016/j.enbuild.2023.113418.

[56] Koukoufikis G, Schockaert H, Paci D, Filippidou F, Caramizaru A, Della Valle N, et al. Energy communities and energy poverty: the role of energy communities in alleviating energy poverty. LU: Publications Office of the European Union; 2023.

[57] Seyfang G, Park JJ, Smith A. A thousand flowers blooming? An examination of community energy in the UK. Energy Policy 2013;61:977–89. https://doi.org/10.1016/j.enpol.2013.06.030.

[58] Heuninckx S, Boveldt G te, Macharis C, Coosemans T. Stakeholder objectives for joining an energy community: Flemish case studies. Energy Policy 2022;162:112808. https://doi.org/10.1016/j.enpol.2022.112808.

[59] Soeiro S, Dias MF. Motivations for Integrating a Renewable Energy Community: Evidence for Spain and Portugal. 2020 17th International Conference on the European Energy Market (EEM), 2020, p. 1–6. https://doi.org/10.1109/EEM49802.2020.9221887.

[60] Dóci G, Vasileiadou E. “Let’s do it ourselves” Individual motivations for investing in renewables at community level. Renewable and Sustainable Energy Reviews 2015;49:41–50. https://doi.org/10.1016/j.rser.2015.04.051.

[61] McKenna R. The double-edged sword of decentralized energy autonomy. Energy Policy 2018;113:747–50. https://doi.org/10.1016/j.enpol.2017.11.033.

[62] McKenna R, Jäger T, Fichtner W. Energieautarkie – ausgewählte Ansätze und Praxiserfahrungen im deutschsprachigen Raum. uwf 2014;22:241–7. https://doi.org/10.1007/s00550-014-0339-y.

[63] McKenna R, Merkel E, Fichtner W. Energy autonomy in residential buildings: A techno-economic model-based analysis of the scale effects. Applied Energy 2017;189:800–15. https://doi.org/10.1016/j.apenergy.2016.03.062.

[64] Rae C, Bradley F. Energy autonomy in sustainable communities—A review of key issues. Renewable and Sustainable Energy Reviews 2012;16:6497–506. https://doi.org/10.1016/j.rser.2012.08.002.

[65] Brummer V. Community energy – benefits and barriers: A comparative literature review of Community Energy in the UK, Germany and the USA, the benefits it provides for society and the barriers it faces.

Renewable and Sustainable Energy Reviews 2018;94:187–96. https://doi.org/10.1016/j.rser.2018.06.013.

[66] Rogers JC, Simmons EA, Convery I, Weatherall A. Public perceptions of opportunities for community-based renewable energy projects. Energy Policy 2008;36:4217–26. https://doi.org/10.1016/j.enpol.2008.07.028.

[67] Madlener R, Crump R. 17 - Grid-friendly clean energy communities and induced intracommunity cash flows through peer-to-peer trading11This chapter is a shortened and updated version of Crump and Madlener (2019), where further details—especially regarding the methodology and assumptions—are available. In: Löbbe S, Sioshansi F, Robinson D, editors. Energy Communities, Academic Press; 2022, p. 277–301. https://doi.org/10.1016/B978-0-323-91135-1.00016-X.

[68] Weckesser T, Dominković DF, Blomgren EMV, Schledorn A, Madsen H. Renewable Energy Communities: Optimal sizing and distribution grid impact of photo-voltaics and battery storage. Applied Energy 2021;301:117408. https://doi.org/10.1016/j.apenergy.2021.117408.

[69] Tostado-Véliz M, Kamel S, Hasanien HM, Turky RA, Jurado F. Optimal energy management of cooperative energy communities considering flexible demand, storage and vehicle-to-grid under uncertainties. Sustainable Cities and Society 2022;84:104019. https://doi.org/10.1016/j.scs.2022.104019.

[70] Mehta P, Tiefenbeck V. Solar PV sharing in urban energy communities: Impact of community configurations on profitability, autonomy and the electric grid. Sustainable Cities and Society 2022;87:104178. https://doi.org/10.1016/j.scs.2022.104178.

[71] Rao BV, Stefan M, Brunnhofer T, Schwalbe R, Karl R, Kupzog F, et al. Optimal capacity management applied to a low voltage distribution grid in a local peer-to-peer energy community. International Journal of Electrical Power & Energy Systems 2022;134:107355. https://doi.org/10.1016/j.ijepes.2021.107355.

[72] Barone G, Buonomano A, Forzano C, Palombo A, Russo G. The role of energy communities in electricity grid balancing: A flexible tool for smart grid power distribution optimization. Renewable and Sustainable Energy Reviews 2023;187:113742. https://doi.org/10.1016/j.rser.2023.113742.

[73] Kerscher S, Koirala A, Arboleya P. Grid-optimal energy community planning from a systems perspective. Renewable and Sustainable

Energy Reviews 2024;199:114485. https://doi.org/10.1016/j.rser.2024.114485.

[74] Sudhoff R, Derzbach R, Schreck S, Thiem S, Niessen S. On the operation and implications of grid-interactive renewable energy communities. Sustainable Energy, Grids and Networks 2024;38:101390. https://doi.org/10.1016/j.segan.2024.101390.

[75] Hashemipour N, Pedrero RA, Granado PC del, Aghaei J. Smart coordination of buildings to incentivise grid flexibility provision: A virtual energy community perspective. Energy and Buildings 2024;310:114078. https://doi.org/10.1016/j.enbuild.2024.114078.

[76] Gržanić M, Capuder T, Zhang N, Huang W. Prosumers as active market participants: A systematic review of evolution of opportunities, models and challenges. Renewable and Sustainable Energy Reviews 2022;154:111859. https://doi.org/10.1016/j.rser.2021.111859.

[77] Irshaid J, Mochizuki J, Schinko T. Challenges to local innovation and implementation of low-carbon energy-transition measures: A tale of two Austrian regions. Energy Policy 2021;156:112432. https://doi.org/10.1016/j.enpol.2021.112432.

[78] Fina B, Fechner H. Transposition of European Guidelines for Energy Communities into Austrian Law: A Comparison and Discussion of Issues and Positive Aspects. Energies 2021;14:3922. https://doi.org/10.3390/en14133922.

[79] Cejka S, Frieden D, Kitzmüller D. Implementation of self-consumption and energy communities in Austria's and EU member states' national law: A perspective on system integration and grid tariffs. CIRED 2021 - The 26th International Conference and Exhibition on Electricity Distribution, , Online Conference: Institution of Engineering and Technology; 2021, p. 3254–8. https://doi.org/10.1049/icp.2021.1526.

[80] European Commission (DG ENER). Energy communities in the clean energy package: best practices and recommendations for implementation. Luxembourg: Publications Office of the European Union; 2020.

[81] Bomberg E, McEwen N. Mobilizing community energy. Energy Policy 2012;51:435–44. https://doi.org/10.1016/j.enpol.2012.08.045.

[82] Reames TG. A community-based approach to low-income residential energy efficiency participation barriers. Local Environment 2016.

[83] Reijnders VMJJ, van der Laan MD, Dijkstra R. Chapter 6 - Energy communities: a Dutch case study. In: Sioshansi F, editor. Behind and Beyond the Meter, Academic Press; 2020, p. 137–55. https://doi.org/10.1016/B978-0-12-819951-0.00006-2.

[84] Energy Communities Repository. Digital tools for energy communities - A short guide. European Commission; 2023.

[85] Gaggero GB, Piserà D, Girdinio P, Silvestro F, Marchese M. Novel Cybersecurity Issues in Smart Energy Communities. 2023 1st International Conference on Advanced Innovations in Smart Cities (ICAISC), Jeddah, Saudi Arabia: IEEE; 2023, p. 1–6. https://doi.org/10.1109/ICAISC56366.2023.10085312.

[86] Kent J, Rowe J, Tomlinson C. Briefing: Information security in community energy. Proceedings of the Institution of Civil Engineers - Energy 2016;169:48–51. https://doi.org/10.1680/jener.15.00023.

[87] Mokarim A, Gaggero GB, Marchese M. Impact Analysis of Cyber Attacks against Energy Communities in Distribution Grids. Electronics 2024;13:1709. https://doi.org/10.3390/electronics13091709.

[88] Bajs D. Coordination Group for Cybersecurity and Critical Infrastructure. Energy Community 2018. https://www.energy-community.org/aboutus/institutions/CyberCG.html (accessed June 14, 2024).

[89] Martilla JA, James JC. Importance-Performance Analysis. Journal of Marketing 1977;41:77–9. https://doi.org/10.1177/002224297704100112.

[90] Pierro M, Cornaro C, Moser D, Perez R, Perez M, Zambotti S, et al. Ground-breaking approach to enabling fully solar renewable energy communities. Renewable Energy 2024; 237:121501. https://doi.org/10.1016/j.renene.2024.121501.

[91] Fischer H, Haas R, Ajanovic A, Radosits F. Energiegemeinschaften – eine Evaluierung bisheriger Erfahrungen und zukünftiger Perspektiven für Österreich. Vienna: Technical University of Vienna; 2024.

[92] Neumann H-M, Stollnberger R, Stortecky S, Etminan G, Schrammel J. Energiegemeinschaften als Bestandteil smarter und nachhaltiger Stadtquartiere. Wien: Wien Energie; 2020.

[93] International Energy Agency. Electricity Grids and Secure Energy Transitions: Enhancing the Foundations of Resilient, Sustainable and Affordable Power Systems. Paris: OECD; 2023. https://doi.org/10.1787/455dd4fb-en.

[94] Savelli I, Morstyn T. Better together: Harnessing social relationships in smart energy communities. Energy Research & Social Science 2021;78:102125. https://doi.org/10.1016/j.erss.2021.102125.

[95] Austrian Energy Agency. Make energy communities fly. Austrian Energy Agency n.d. https://shares-project.eu/ (accessed August 9, 2024).

[96] Energy Saving Trust. CARES. Energy Saving Trust n.d. https://energysavingtrust.org.uk/programme/community-and-renewable-energy-scheme/ (accessed August 9, 2024).
[97] Énergie Partagée. Le Mouvement Énergie Partagée. Énergie Partagée n.d. https://energie-partagee.org/decouvrir/le-mouvement/ (accessed August 9, 2024).

3

Reşiţa, Romania – A Case Study in Urban Decarbonization and Climate Adaptation Through SECAP

C. Vasilica

Denkstatt Romania, Bucharest Tower Center Building, Romania
E-mail: Catalina.Vasilica@ro.ey.com

Abstract

Urban regions are responsible for significant global greenhouse gas emissions, making them crucial targets for climate change mitigation efforts. Reşiţa, a city in western Romania, has shown a strong interest in urban decarbonization and climate adaptation efforts, developing a sustainable energy and climate action plan to address the challenges inquired by climate change. This case study examines the strategies and initiatives undertaken by the city to reduce its carbon footprint and enhance resilience. The research explores the barriers and opportunities encountered in the implementation process, as well as the role of public policy and stakeholder engagement in driving these transformative changes.

Keywords: Sustainable energy and climate action plans, urban planning, energy and climate targets, CO_2 emissions.

3.1 Introduction

The global urgency to address climate change has prompted cities worldwide to develop and implement strategies to reduce their greenhouse gas (GHG)

DOI: 10.1201/9788743806851-3

emissions and build resilience to the adverse effects of a changing climate. As major hubs of economic activity and population density, urban areas account for a significant share of global emissions, making them critical components of any comprehensive decarbonization effort [1, 2].

In response to this urgency, the Paris Agreement established a framework for countries to commit to reducing their GHG emissions through Nationally Determined Contributions (NDCs). Building on this global commitment, the European Union (EU) launched a set of climate actions under the European Green Deal in 2019, aiming to decouple the European economy from energy and resource use and achieve carbon neutrality by 2050. Key legislative frameworks supporting these goals include the European Climate Law [3] and the European Climate Pact [4].

To align with these EU targets, member states were required to submit their National Energy and Climate Plans (NECPs) for the period 2021–2030. These plans outline each country's roadmap for transitioning to a low-carbon economy, mirroring the NDCs' role in the Paris Agreement.

At a more localized level, cities have also committed to sustainable development. Since 2008, the Covenant of Mayors for Climate and Energy (CoM) [5] has brought together local governments that voluntarily pledge to meet and exceed EU climate and energy targets. These cities aim to reduce their GHG emissions and ensure access to secure, sustainable, and affordable energy for all residents [6]. Similar to the NECPs for countries, cities participating in the CoM are required to submit a Sustainable Energy and Climate Action Plan (SECAP).

The city of Reşiţa, located in western Romania, has recognized its responsibility in this global effort. Reşiţa has undertaken a joint initiative to transition to a low-carbon and climate-resilient future. The city of Reşiţa adhered to the Covenant of Mayors commitment in 2009, developing a Sustainable Energy Action Plan (SEAP) [7] to address energy challenges. In 2024, Reşiţa renewed its commitment by aligning with the most ambitious CoM initiatives, transitioning to a SECAP to further enhance its climate resilience and sustainability efforts.

Developed in partnership with the Covenant of Mayors for Climate and Energy, Reşiţa's SECAP serves as the foundation for the city's climate action strategy [1, 8].

Hence, this case study examines Reşiţa's pioneering efforts in urban decarbonization and climate adaptation, highlighting the city's adoption of the sustainable energy and climate action plan as a framework for addressing the complex challenges of climate change.

3.2 Methodology and Approach

3.2.1 Policies and instruments for addressing climate change in urban areas

The European Union has positioned the fight against climate change at the forefront of its agenda, setting ambitious targets to reduce greenhouse gas emissions in the medium and long term. By 2030, the EU aims to decrease emissions by 55% as a milestone toward the more distant objective of achieving net-zero emissions by 2050 [9]. This commitment reflects a concerted effort to transition toward a sustainable and environmentally responsible economy.

Building upon these objectives, the "2030 Climate and Energy Framework" sets rigorous targets for 2030, including a decrease in emissions, an increase in the usage of renewable energy sources, and an enhancement in energy efficiency. These objectives are in line with the "Energy Roadmap 2050," which aims to transition to a low-carbon economy by achieving an 80% cut in domestic greenhouse gases by 2050. The "2030 Framework" is also fundamental in fulfilling the goals of the Paris Agreement, which aims to keep the increase in global average temperature well below 2°C above pre-industrial levels, pursuing efforts to limit the temperature increase even further to 1.5°C. The agreement includes mechanisms for scaling up ambitions, a financial framework to promote solidarity, and supports adaptation and resilience to climate change, motivating a wide array of actors to intensify their efforts in emission reduction and resilience building [3].

At the policy level, the Reşiţa SECAP is embedded within a broader framework of national and regional climate and energy strategies. This integration ensures that local actions are coherent with and contribute to overarching climate and energy goals, thereby enhancing the effectiveness and impact of the initiatives undertaken.

The process involved in developing the climate and action plan following the key steps [10]:

- Commitment: Municipalities voluntarily commit to the Covenant of Mayors initiative, aiming to reduce CO_2 emissions by a specified target year.
- Baseline Emission Inventory (BEI): Municipalities establish a baseline of current CO_2 emissions from sectors such as residential, tertiary (commercial), transport, industry, and others.
- Risk and vulnerability assessment (RVA): Assessing the risks and vulnerabilities of the local area to climate change impacts.

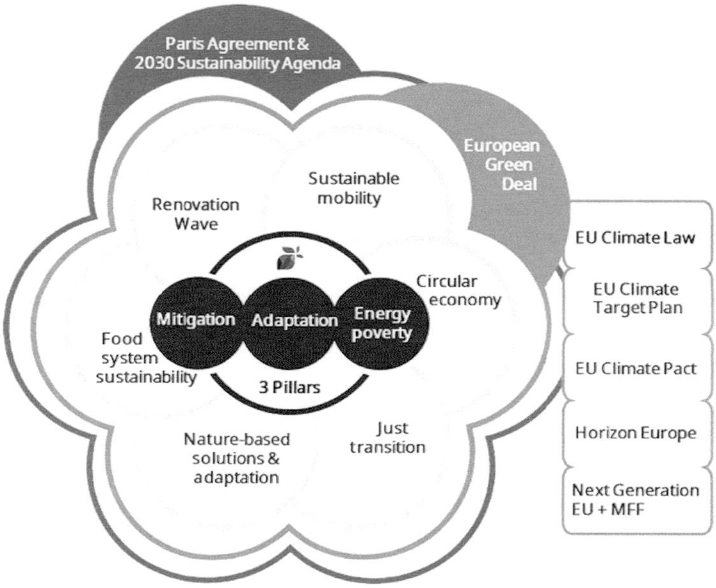

Figure 3.1 Overview of the main steps for a municipality to join the Covenant of Mayors for Sustainable Climate and Energy Action Plan [5].

- Target setting: Goals for emission reduction and climate resilience are set, informed by initial assessments.
- Developing the action plan: The SECAP details measures for emission mitigation and climate adaptation, with clear roles, timelines, and financial projections.
- Implementation and monitoring: Actions are implemented, progress is monitored, and the plan is adjusted as needed.
- Reporting: Regular updates are provided to the Covenant of Mayors, promoting transparency and best practice exchange.
- Stakeholder involvement: Engagement with community members and organizations is essential for building support and enhancing the initiative's capacity.

3.2.2 Policy and regulatory considerations – integrating with SDG and IPPC

The Sustainable Development Goals (SDGs) and IPCC Cities present matchless opportunities for the transformation of urban environments. However, to overcome the challenges presented by economic, cultural, and political factors, there is a necessity for decisive and comprehensive measures [11].

The SDGs and targets (Table 3.1) are part of a universal call to action to end poverty, protect the planet, and ensure that all people enjoy peace and prosperity by 2030, to which Reşiţa's SECAP has aligned (Figure 3.2). The

Table 3.1 SDGs targets aligned to climate adaptation and mitigation measures proposed through SECAP [12].

SDG 7: Affordable and Clean Energy	
Target 7.1	By 2030, ensure universal access to affordable, reliable, and modern energy services.
Target 7.2	By 2030, increase substantially the share of renewable energy in the global energy mix.
SDG 9: Industry, Innovation, and Infrastructure	
Target 9.1	Develop quality, reliable, sustainable, and resilient infrastructure, including regional and transborder infrastructure, to support economic development and human well-being, with a focus on affordable and equitable access for all.
SDG 11: Sustainable Cities and Communities	
Target 11.2	By 2030, provide access to safe, affordable, accessible, and sustainable transport systems for all, improving road safety, notably by expanding public transport, with special attention to the needs of those in vulnerable situations, women, children, persons with disabilities, and older persons.
Target 11.4	Strengthen efforts to protect and safeguard the world's cultural and natural heritage.
Target 11.6	By 2030, reduce the adverse per capita environmental impact of cities, including by paying special attention to air quality and municipal and other waste management.
Target 11.7	By 2030, provide universal access to safe, inclusive, and accessible green and public spaces, for women and children, older persons, and persons with disabilities.
SDG 12: Responsible Consumption and Production	
Target 12.5	By 2030, substantially reduce waste generation through prevention, reduction, recycling, and reuse.
Target 12.8	By 2030, ensure that people everywhere have the relevant information and awareness for sustainable development and lifestyles in harmony with nature.
SDG 13: Climate Action	
Target 13.1	Strengthen resilience and adaptive capacity to climate-related hazards and natural disasters in all countries.
Target 13.2	Integrate climate change measures into national policies, strategies, and planning.
Target 13.3	Improve education, awareness-raising, and human and institutional capacity on climate change mitigation, adaptation, impact reduction, and early warning.

Figure 3.2 Integrating SDGs with Reşiţa's sustainable energy and climate action plan [12].

plan outlines a comprehensive strategy that includes reducing greenhouse gas emissions, increasing energy efficiency, and promoting RES. This approach not only addresses climate change mitigation (SDG 13) but also promotes sustainable cities and communities (SDG 11) by improving air quality, enhancing public transportation, and creating green spaces.

Furthermore, Reşiţa's SECAP integrates aspects of responsible consumption and production (SDG 12) through waste management and circular economy initiatives. The city's efforts to provide affordable and clean energy (SDG 7) are exemplified by investments in photovoltaic (PV) power projects, contributing to a diversified and sustainable energy mix. A highlight of these efforts is the establishment of an 8.6-megawatt (MW) solar photovoltaic (PV) facility, which will fully meet the renewable energy requirements for public transportation systems (including trams and electric buses), as well as for public structures (such as schools, kindergartens, cinemas, and government offices), and street lighting.

The city's holistic approach to urban development showcases the potential for local actions to contribute to global sustainability and climate objectives.

3.3 The Reşiţa, Romania Case Study: Results and Discussions

Reşiţa (Figure 3.3), situated in western Romania's Banat region, is the administrative hub of Caraş-Severin County, with a population of around 84,564 in

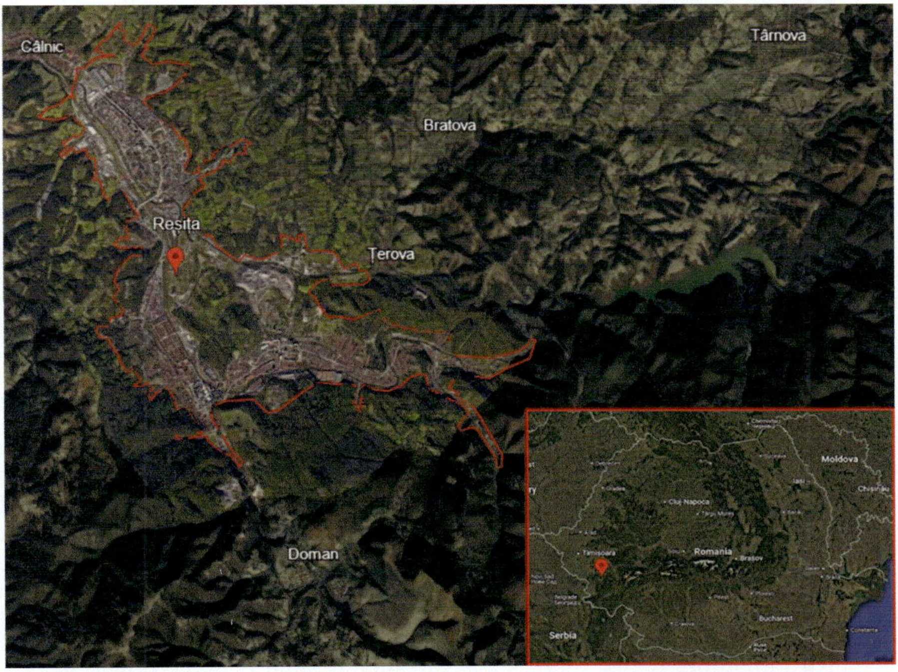

Figure 3.3 Reşiţa, Romania – Location of the case study (Google maps).

2020. Known for its strong industrial roots, Reşiţa has long been recognized as one of Romania's major industrial centers, with significant contributions to the steel industry (including cast iron, steel, and laminates), machine construction (such as electric and diesel engines, oil and steel equipment, chemical apparatus, and rolling stock), and the chemical sector (notably wood chemicalization and coke production). It remains an important hub for steel processing and machine building. The city's industrial backbone has been supported by key factories like Combinatul Siderurgic Reşita (CSR), now known as TMK-Reşita, and Uzina Constructoare de Maşini Reşita (UCMR), which have sustained Reşiţa for over 300 years. The first factories were established in 1771 under the reign of Empress Maria Theresa, with the inaugural blast furnaces being lit on July 3, 1771, at the site now occupied by SC TMK Reşita SA. Therefore, the city plays a crucial role in the region's economy.

Geographically, Reşiţa is advantageously positioned near the Semenic Mountains, offering easy access to natural attractions and outdoor activities.

Table 3.2 Reşiţa's SECAP 2030 and 2050 objectives

	Base year (2008)	Objective (2030)	Objective (2050)
40% reduction GHG emissions (Tons CO_2eq) 80% reduction GHG emission (Tons CO_2eq) Adaptation to climate change Access to secure, sustainable, and affordable energy	216783	86713	173426

In 2022, Reşiţa has become the first city in Romania to join the UNESCO Global Network "Learning Cities" [13].

The SECAP case study of Reşiţa, Romania (Table 3.2) provides valuable insights into the implementation of sustainable energy initiatives at the local level. Reşiţa has been actively engaged in efforts to transition toward a low-carbon future, with a focus on improving energy efficiency, increasing the use of renewable energy, and enhancing overall urban sustainability.

One of the key findings from the case study is the importance of aligning local sustainability plans, such as the Sustainable Energy and Climate Action Plan, with national and European decarbonization strategies. This alignment is crucial to ensure that local actions are contributing to the achievement of higher-level targets and that cities are receiving the necessary support and resources to implement their plans effectively.

Another notable aspect of Reşiţa's approach is the emphasis on holistic urban planning, which integrates energy, climate adaptation, and resilience considerations. By adopting a comprehensive perspective, Reşiţa has been able to address multiple urban challenges simultaneously, enhancing the overall sustainability and climate readiness of the city [14].

3.3.1 Baseline energy consumption analysis

Using the baseline year of 2008, Reşiţa's SECAP identifies the city's total energy consumption and associated greenhouse gas emissions, establishing a reference point for tracking progress.

The sectors covered by the emission inventories within the Sustainable Energy and Climate Action Plan are structured according to the most recent version of the Covenant of Mayors (Figure 3.4).

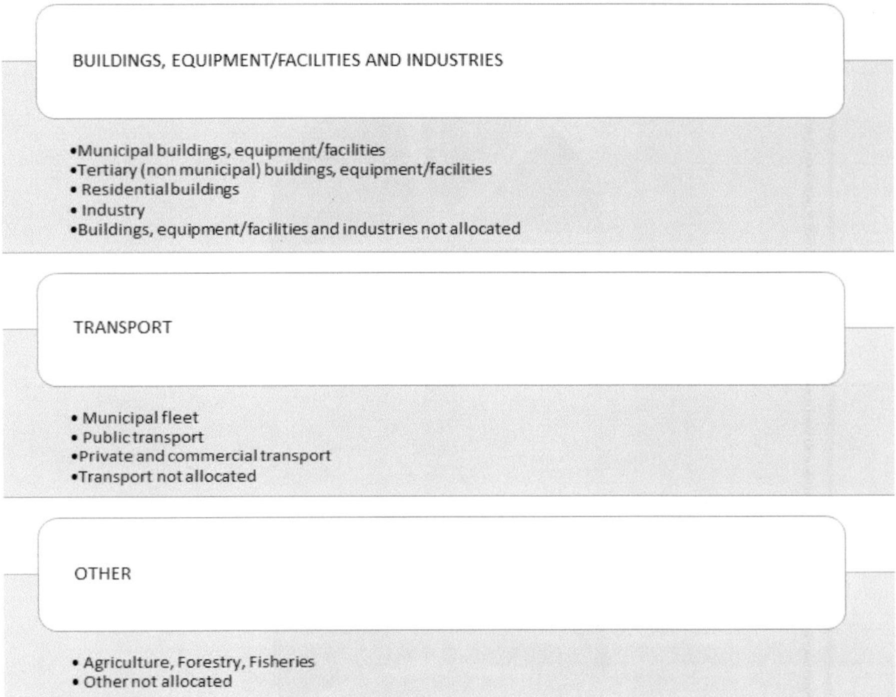

Figure 3.4 Sectors covered by the emission.

3.3.2 Distribution of energy consumption by sectors of the Municipality of Reşiţa, Romania

Analyzing energy consumption in the different sectors included in the SECAP (Figure 3.5, Figure 3.6) allows us to identify priorities and take concrete measures to optimize energy consumption. Here are some significant findings:

- *The buildings sector* dominates energy consumption (accounting for about 75% of total consumption in BEI and 72% in 2020). Here, *residential buildings* account for the largest share, at 47%. This sector should be a priority as there is potential for significant improvements in energy efficiency in housing.
- *The tertiary sector and municipal buildings* account for a significant share of consumption (around 28% in total). In this context, it is important to identify solutions to reduce energy consumption in these

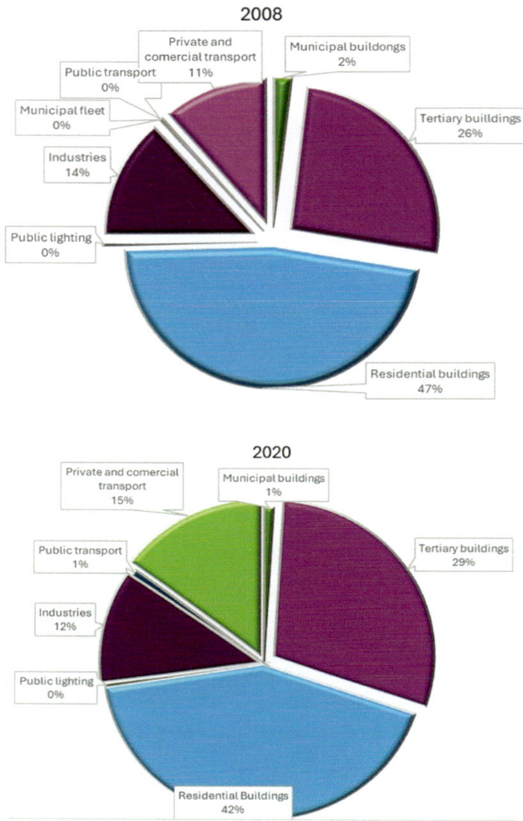

Figure 3.5 Distribution of energy consumption by sectors, Municipality of Reşiţa – 2008, 2020 [14].

types of buildings and to promote renewable energy and efficient technologies.

– *The industrial sector accounts for* a significant share of around 14% of total energy consumption. This sector can benefit from investments in energy efficiency technologies, process optimization, and the use of cleaner energy sources.
– *Private and commercial transport* accounts for about 11% of total energy consumption. Improving vehicle efficiency and promoting sustainable transport can make a significant contribution to reducing this consumption.

The increase in energy consumption in the tertiary sector can be attributed to the expansion and opening of new commercial outlets, such as shopping centers. This expansion brings several implications that contribute to increased energy demand, including:

- Increased commercial floor space: The addition of new commercial spaces requires more energy for heating, cooling, lighting, and other operational needs.
- Increased use of technology: Modern commercial outlets often incorporate advanced technologies, which, while improving efficiency and customer experience, also contribute to higher energy consumption.
- Transport and logistics: The logistics associated with stocking and maintaining commercial outlets involve transportation and storage, both of which require significant energy inputs.
- Air conditioning and ventilation: Larger commercial spaces necessitate extensive air conditioning and ventilation systems to maintain a comfortable environment for shoppers and staff.
- Lighting and commercial equipment: The need for extensive lighting and the use of various commercial equipment, such as escalators, elevators, and digital displays, further drive up energy consumption.
- Extended opening hours: Longer operating hours mean that energy-consuming systems and equipment are in use for extended periods, contributing to higher overall energy demand.

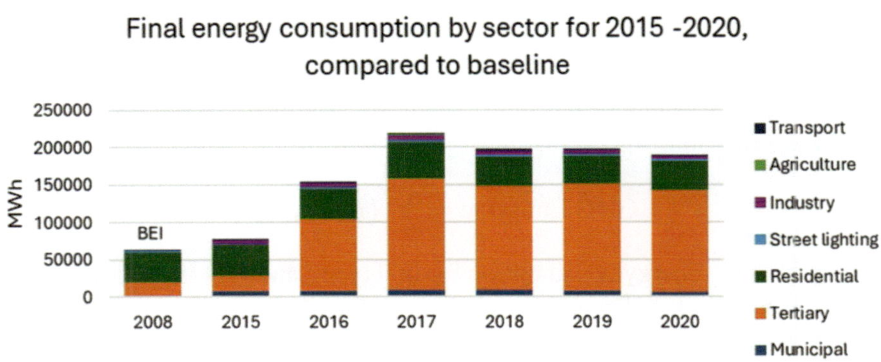

Figure 3.6 Final energy consumption by sector compared to baseline of 2008, [12],

Therefore, the expansion of commercial outlets, particularly shopping centers, is a significant factor in the increasing energy consumption observed in the tertiary sector (see Figure 3.6).

3.3.3 Mitigation measures: Energy efficiency in buildings

One of the key mitigation strategies outlined in Reşiţa's SECAP is the improvement of energy efficiency in both residential and non-residential buildings.

The plan proposes measures such as thermal insulation upgrades, window replacements, and the installation of high-efficiency heating and cooling systems. These retrofits are expected to deliver substantial reductions in energy consumption and associated emissions [6].

Based on the analysis of Reşiţa's energy consumption patterns, the residential sector emerges as the largest consumer, accounting for 72% of the city's total energy use [14].

The commercial and institutional buildings sectors also represent significant shares, highlighting the importance of targeting these areas for energy efficiency improvements.

The SECAP's mitigation strategies, if fully implemented, are expected to result in a reduction in energy use and a decrease in greenhouse gas emissions by 2030, compared to the baseline year.

This projected impact demonstrates the significant potential for urban decarbonization through targeted initiatives, such as building energy retrofits, renewable energy deployment, and sustainable transportation solutions.

3.3.4 Emission assessment and sectoral distribution in Reşiţa

The baseline greenhouse gas emissions for Reşiţa were calculated using the following formula:

$$\text{GHG Emission} = \sum(E_i \times EF_i),$$

where E_i is the energy consumption for fuel type i and EF_i is the corresponding emission factor. The CO_2 emissions resulting from the monitored consumption levels were calculated using the conversion factors recommended by the Intergovernmental Panel on Climate Change (IPCC). Additionally, for municipal waste, the life cycle assessment (LCA) factor was considered, and for electricity, the national factor according to the to the 2016 IPCC report. IPCC report (Figure 3.7).

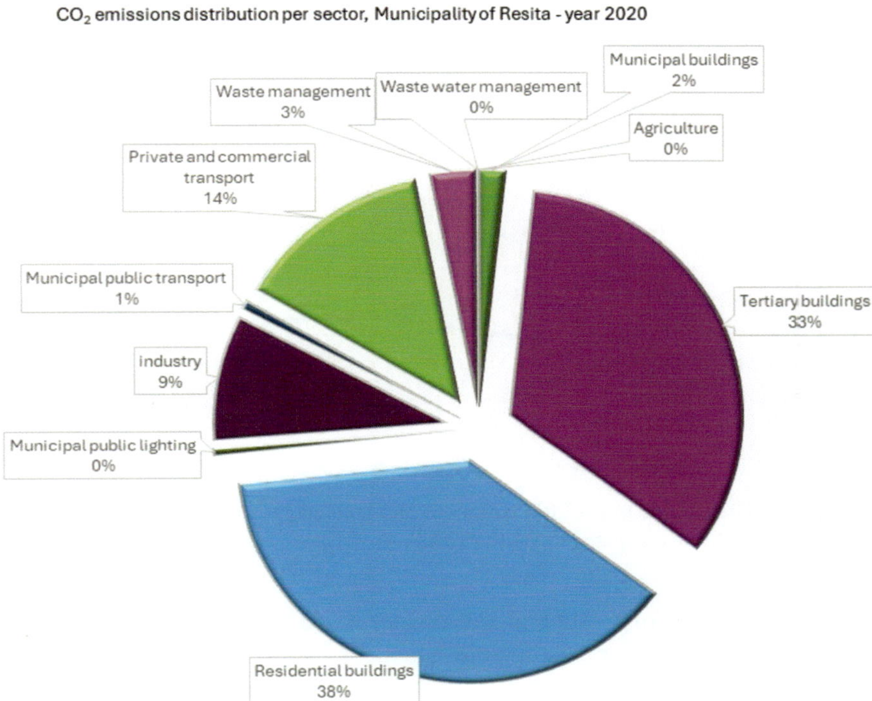

Figure 3.7 Distribution of CO_2 emissions by sectors in 2020, [12].

3.3.5 Impact on carbon emission reduction

The analysis of CO_2 emissions in the Municipality of Reşiţa for the year 2020 offers a comprehensive perspective on the environmental impact across various sectors. Residential buildings account for 38% of CO_2 emissions, highlighting the need for targeted interventions in this domain. Municipal street lighting, on the other hand, exhibits minimal CO_2 emissions, indicative of the successful implementation of energy-efficient lighting solutions, such as the widespread adoption of LED technology.

The industrial sector, contributing 9% to the total emissions, presents significant opportunities for the integration of sustainable production practices aimed at mitigating environmental impact. The transport sector, encompassing both public (1%) and private/commercial (14%) transportation, emerges as a notable source of emissions, thereby highlighting the urgency of transitioning to low-emission transport solutions.

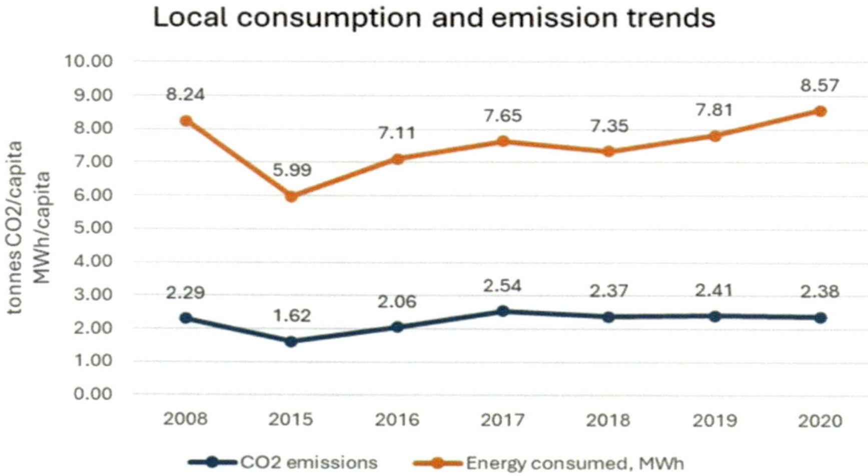

Figure 3.8 Greenhouse gas (GHG) emissions and final energy consumption per capita for the period 2015–2020 compared to baseline 2008.

In the domains of waste and wastewater management, the analysis highlights the necessity for sustainable practices, with a particular focus on recycling and resource efficiency. Municipal and tertiary buildings contribute 2% and 33% respectively to the overall CO_2 emissions, highlighting the critical importance of deploying energy-efficient technologies and practices within public and commercial infrastructures.

The impact of energy consumption on the local carbon footprint in the Municipality of Reşiţa is quantified by emissions expressed in tons of CO_2 per year per capita (Figure 3.8). A general decreasing trend in CO_2 emissions was observed in 2015, followed by a relative increase in subsequent years. By implementing community-level measures, the per capita carbon footprint in 2020 was reduced to 2.38 tons of CO_2 per year [14].

These developments underscore the necessity for a comprehensive analysis of the factors influencing these changes to identify potential measures and policies that could further contribute to reducing CO_2 emissions and optimizing energy consumption in the future.

This analysis provides a holistic overview of the emission sources within the municipality, thereby offering clear directives for the formulation of sustainable and energy-efficient strategies essential for achieving the carbon footprint reduction targets in the Municipality of Reşiţa.

The analysis reveals that the buildings sector, particularly residential and commercial buildings, accounts for the largest share of energy use and emissions, underscoring the importance of targeting this sector for mitigation efforts.

3.3.6 Renewable energy integration strategies

Reşiţa's SECAP emphasizes the integration of renewable energy sources as a key strategy to reduce the city's carbon footprint.

One of the primary initiatives is the installation of rooftop photovoltaic systems on municipal buildings and private residences to increase the share of renewable electricity generation.

Complementing the energy efficiency measures, these renewable energy initiatives are expected to significantly contribute to Reşiţa's decarbonization efforts by offsetting the consumption of grid electricity, which is still largely generated from fossil fuels in the region.

From the mitigation actions, the SECAP outlines ambitious targets for renewable energy integration, aiming to increase the share of renewable sources, primarily photovoltaic panels, to account for a substantial proportion of the city's total energy consumption by 2030.

3.3.7 Risk and vulnerability assessment used to prioritize areas for action within the SECAP

An important step in the SECAP process was the completion of a comprehensive risk and vulnerability assessment.

This assessment examined the city's exposure and sensitivity to the projected impacts of climate change, including increased temperatures, extreme weather events, and water scarcity.

The assessment integrated both quantitative and qualitative data to identify the sectors and population groups most vulnerable to these climate risks [8, 15].

The findings of the risk assessment then informed the prioritization of mitigation and adaptation measures within the SECAP, ensuring that the plan's objectives and actions were tailored to the specific needs and challenges facing Reşiţa.

Reşiţa's SECAP process began with a thorough risk and vulnerability assessment to identify the most pressing climate-related threats facing the city and its population.

This comprehensive analysis examined a range of climate hazards, including rising temperatures, more frequent and intense extreme weather events, and the potential for flooding and drought [11, 16] (Table 3.3).

Table 3.3 Identified risks in Reşiţa.

Type of climate hazard	Current risks	Anticipated risks			Risk indicators
	Current level of hazard risk	**Expected change in intensity**	**Expected change in frequency**	**Time horizon**	
Extreme heat	High	Increase	Increase	Short term	Annual number of tropical days ($T_{max} \geq 30°C$); frequency and duration of heatwaves (days/year); annual maximum recorded temperature
Extreme cold	Low	Increase	No change	Short term	Annual number of frost days ($T_{min} < 0°C$); duration of cold spells (days/year)
Extreme precipitation	High	Increase	Increase	Short term	Maximum daily precipitation (mm/24h); total precipitation amount during extreme events (mm); sediment load (kg/m^2 per 24h)
Flooding	High	Increase	Increase	Short term	Number of flood events/ year; total inundated area (km^2); number of affected settlements; estimated flood damage (€)
Drought	Moderate	Increase	Increase	Short term	Number of drought events/year; crop yield reduction (%)
Storms	High	Increase	Increase	Short term	Number of storm events/year; maximum wind gust speed (m/s); annual storm-related damage (€)
Heavy snowfall	Low	Increase	Increase	Short term	Number of heavy snowfall events/ year; total snowfall amount (cm/event); duration (days/year); associated damage (€)
Forest fires, vegetation fires	Low	Increase	Increase	Medium term	Number of wildfire events/ year; burned area (ha); estimated damage costs (€)
Landslides	Moderate	Increase	Increase	Short term	Number of landslide events/year; affected area (ha); damage costs (€)
Biological hazards	Low	Uncertain	Increase	Long term	Number of reported illiness cases/year; crop yield loss (%) due to pests or diseases; economic loss (€) from agricultural damage

Table 3.4 Mitigation actions.

Mitigation sectors	Number of actions included in the plan
Municipal buildings, equipment/facilities	37
Tertiary (non-municipal) buildings, equipment/facilities	3
Residential buildings	20
Transport	22
Waste	6
Local electricity production	6
Others	18
Total	**112**

The assessment also considered the vulnerabilities of Reşiţa's built environment, critical infrastructure, and socioeconomic systems to these climate impacts [17].

The findings of this risk and vulnerability assessment were then used to guide the development of Reşiţa's emissions reduction targets and adaptation strategies, ensuring that the SECAP addressed the city's most pressing climate-related challenges [11, 18]

3.3.8 Action plan

Reşiţa's SECAP outlines a detailed roadmap for reducing greenhouse gas emissions and enhancing climate resilience through a series of mitigation and adaptation measures.

The mitigation component of the plan focuses on seven key action areas, encompassing a total of 112 specific measures. These measures span across sectors such as energy efficiency in buildings, increased use of renewable energy, sustainable transport, and waste management.

The adaptation component of the SECAP identifies 16 priority actions (Table 3.5) to enhance the city's preparedness and responsiveness to climate change impacts. These initiatives address areas such as water resource management, urban greening, and emergency planning.

By integrating both mitigation and adaptation strategies, Reşiţa's SECAP recognizes the interdependence of these two pillars of climate action and seeks to pursue a holistic approach to building a sustainable, resilient city.

Under the framework of SECAP, the city has implemented a comprehensive set of initiatives designed to lower greenhouse gas emissions, enhance energy efficiency, and strengthen the adoption of renewable energy. These initiatives are pivotal to the city's efforts to decarbonize its urban landscape, a crucial step in mitigating climate change impacts.

Table 3.5 Adaptation actions.

Adaptation sectors	Number of actions included in the plan
Transport	2
Waste	5
Land use planning	5
Environment and Biodiversity	1
Health	1
Civil protection and emergency	1
Other	1
Total	**16**

Reşiţa's strategy for urban decarbonization includes a series of infrastructure improvements. These improvements encompass the modernization of buildings for better energy efficiency, and a shift toward eco-friendly transportation options. The plan proposes the gradual transition to electric buses in the municipal public transport fleet, leveraging the city's commitment to sustainable mobility and cleaner air. The shift toward sustainable mobility will help mitigate emissions from the transport sectors, which significantly contributes to the city's carbon footprint [19].

Alongside, the city is addressing climate adaptation by strengthening its urban environment against the repercussions of climate change. Strategies include bolstering infrastructure resilience and efficiently managing the city's water resources. Reşiţa's approach to climate adaptation involves creating green urban areas to mitigate the heat island effect, establishing systems for rainwater collection to avert flooding, and strengthening essential infrastructure to endure severe weather conditions.

3.3.9 Creating synergy among stakeholders

The case study highlights the pivotal role of community engagement and stakeholder collaboration in the development and implementation of the SECAP.

Achieving the ambitious goals outlined in Reşiţa's SECAP requires the collaboration and engagement of multiple stakeholders, including the local administration, private sector, and civil society. In the scope of SECAP, the city has fostered a collaborative governmental approach involving a diverse range of actors to catalyze mitigation and climate actions through workshop sessions and data collection. By leveraging the strengths and resources of all stakeholders, Reşiţa aims to create a resilient and sustainable urban environment for its residents for long-term systemic transformation [20].

3.3.10 Monitoring and evaluation frameworks

To ensure the success of the plan, Reşiţa relies on thorough data collection and analysis, which enables the city to monitor its progress and make evidence-based decisions. The city has set precise benchmarks for reducing emissions, increasing the use of renewable energy, and enhancing energy conservation. The incorporation of monitoring and evaluation systems is essential for SECAP to evaluate the impact of the measures in place and to fine-tune strategies as needed.

Hence, monitoring and reporting progress will be reported by Reşiţa to Covenant of Mayors according to the CoM methodology [10].

3.4 Conclusion

Climate change adaptation and mitigation in urban areas require a multifaced approach that integrates various policy tools and planning frameworks [1, 21]. This study highlights the importance of a comprehensive and integrated approach to urban decarbonization and climate adaptation. One of the key strategies outlined in the SECAP is the integration of renewable energy sources, particularly installing photovoltaic systems.

The plan aims to significantly increase the city's renewable energy capacity, which will contribute to reducing reliance on fossil fuels and lowering greenhouse gas emissions. The SECAP also includes initiatives to enhance energy efficiency in the building sector, such as thermal insulation upgrades and the installation of high-efficiency heating and cooling systems. These measures are expected to deliver substantial reductions in energy consumption and associated emissions. Furthermore, the plan addresses the transportation sector, with the integration of electric buses into the public transport system.

The Reşiţa case study demonstrates that implementing the initiatives from the action plan is expected to deliver a significant reduction in Reşiţa's overall carbon emissions over the next decade. Estimates suggest that the full implementation of the SECAP's mitigation and climate adaptation measures could lead to a significant percentage decrease in the city's greenhouse gas emissions by 2030, compared to the baseline year.

In conclusion, the SECAP for Reşiţa, Romania, represents a proactive step toward urban decarbonization and climate adaptation. By focusing on sustainable energy practices and resilience-building measures, the city can work toward a greener, more sustainable future that not only contributes to global climate goals but also enhances the quality of life for its residents. Such a plan needs to be dynamic, data-driven, and inclusive to ensure its success and sustainability over the long term.

Acknowledgments

This project has been developed in the framework of the PED Program, which is implemented by the Joint Programming Initiative Urban Europe and SET Plan Action 3.2. The Austrian part is supported by the Austrian Ministry of Climate Action, Environment, Energy, Mobility, Innovation, and Technology (BMK); the Romanian part is supported by a grant of the Ministry of Research, Innovation, and Digitization CNCS/CCCDI – UEFISCDI, project number PED-JPI-SIMPLY POSITIVE, contracts number 325/2022 and 326/2022, within PNCDI III; the Dutch part is supported by the RVO (the Netherlands Enterprise Agency), reference number ERANETPED-02767306; and the Italian part is supported by a grant of the Ministry of Education and Merit – Department for Higher Education and Research, project number PED_00042, from the Fund for Investment in Scientific and Technological Research (FIRST/FAR) and/or Special Ac-counting Account no. 5944.

References

[1] W. N. M. W. M. Rani, K. H. Kamarudin, K. A. Razak, and Z. M. Asmawi, 'Climate Change Adaptation and Disaster Risk Reduction in Urban Development Plans for Resilient Cities,' IOP Conference Series: Earth and Environmental Science, vol. 409, p. 012024, Jan. 2020. [Online]. Available: https://doi.org/10.1088/1755-1315/409/1/012024.

[2] World Bank, 'The Low Carbon City Development Program (LCCDP) Guidebook. A systems approach to low carbon development in cities', 2014. Available at: https://openknowledge.worldbank.org/server/api/core/bitstreams/1bad95c5-b831-5887-8eb8-d5962cf3d21e/content(Accessed:14February2025).

[3] European Climate Law, 2021. [Online]. Available: https://eur-lex.europa.eu/legal-content/EN/TXT/?uri=CELEX%3A32021R1119&qid=1730241634536

[4] European Climate Pact. [Online]. Available: https://climate-pact.europa.eu/index_en

[5] European Commission. [Online]. Available: https://eu-mayors.ec.europa.eu/en/resources/reporting
[6] I. Munoz, P. H.-I.-G., 'How can cities effectively contribute towards decarbonization targets? A downscaling method to assess the alignment of local energy plans with national strategies,' Elsevier, 2023.
[7] Primaria Reşiţa, 'Raport Strategic şi Plan de Acţiune Privind Energia Durabilă – Municipiul Reşiţa,' 2016.
[8] C. Saavedra, 'Assessing Resilience to Climate Change in US Cities,' Urban Studies Research, 2012.
[9] European Green Deal, 2024. [Online]. Available: https://www.consilium.europa.eu/en/policies/green-deal/
[10] P. Bertoldi, Ed., Guidebook: How to Develop a Sustainable Energy and Climate Action Plan (SECAP), European Commission, Joint Research Centre, Luxembourg: Publications Office of the European Union, 2018. [Online]. Available: https://doi.org/10.2760/223399
[11] P. Romero-Lankao et al., 'Urban transformative potential in a changing climate,' Nature Climate Change, vol. 8, no. 9, pp. 754–756, Sep. 2018. [Online]. Available: https://doi.org/10.1038/s41558-018-0264-0
[12] United Nations. [Online]. Available: https://sdgs.un.org/goals
[13] Reşiţa - Green City, 2023. [Online]. Available: https://www.primariaresita.ro/portal/cs/resita/portal.nsf/AllByUNID/7EEE0F5EDB3B1E1DC22589DA00489050/\protect\T1\textdollarFILE/Resita%20Green%20Bonds%20Framework_en.pdf
[14] Integrated SECAP Report for Reşiţa Municipality, Simply Positive Project, 2024. [Online]. Available: http://simplypositive.eu/wp-content/uploads/2024/12/D4.5-SECAP-Resita-Municipality_upload-on-CoM-platform.pdf
[15] N. Chen, X. Tang, and W. Liu, 'Urban Disaster Risk Prevention and Mitigation Strategies from the Perspective of Climate Resilience,' Wireless Communications and Mobile Computing, vol. 2022, Article ID 4907084, 13 pages, 2022. [Online]. Available: https://doi.org/10.1155/2022/4907084.
[16] A. Revi et al., 'Towards transformative adaptation in cities: the IPCC's Fifth Assessment,' Environment and Urbanization, vol. 26, no. 1, pp. 11–28, Apr. 2014. [Online]. Available: https://doi.org/10.1177/0956247814523539
[17] E. B. deSouza Leao, J. C., 'Recife: A Climate Action Profile,' Elsevier, 2021.

[18] X. He, M. Lin, T.-L. Chen, B. Liu, P.-C. Tseng, W. Cao, and P.-C. Chiang, 'Implementation Plan for Low-carbon Resilient City Towards Sustainable Development Goals: Challenges and Perspectives,' Aerosol and Air Quality Research, vol. 20, no. 3, pp. 444–464, Mar. 2020. [Online]. Available: https://doi.org/10.4209/aaqr.2019.11.0568.

[19] R. Sancha Llamosí, E. Coll Aliaga, M. J. Porres De La Haza, V. Lerma Arce, and E. Lorenzo-Sáez, 'Analysis of Decarbonization Potential in Mobility Sector with High Spatial Resolution: Study Case of the Metropolitan Area of Valencia (Spain),' Applied Sciences, vol. 14, no. 16, p. 7340, Aug. 2024. [Online]. Available: https://doi.org/10.3390/ap p14167340.

[20] S. Linton, A. Clarke, and L. Tozer, 'Strategies and Governance for Implementing Deep Decarbonization Plans at the Local Level,' Sustainability, vol. 13, no. 1, p. 154, Jan. 2021. [Online]. Available: https: //doi.org/10.3390/su13010154

[21] M. Ellena, M. Soriani, and M. Breil, 'The Heat-Health Nexus in the Urban Context: A Systematic Literature Review Exploring the Socio-Economic Vulnerabilities and Built Environment Characteristics,' Urban Climate, vol. 34, p. 100676, Dec. 2020. [Online]. Available: https://doi.org/10.1016/j.uclim.2020.100676

4

Digital Services for Energy Transition: Enhancing Quality of Life and Delaying Electricity Grid Retrofitting

S. Fava[1], A. Vaitulevice[2], G. Roccella[1], and A. Azarmi Khah[1]

[1]Planet Smart City, Corso Stati Uniti, Italy
[2]Ignitis Group, Lithuania
E-mail: s.fava@planetsmartcity.com; astasvir@gmail.com; g.roccella@planetsmartcity.com; a.khah@planetsmartcity.com

Abstract

In the context of European policies aimed at promoting energy efficiency, digital energy services play a crucial role in integrating all stakeholders in the value chain, from production to the end consumer. These services foster improved comfort, lower costs, and reduced environmental impacts. They also empower consumers, raising awareness and encouraging them to move from passive users to active participants in energy systems. This increased consumer awareness contributes to the stability of energy distribution networks and can even delay the need for costly grid retrofitting interventions.

One of the key outcomes of the InterConnect project, funded by the European Commission, has been the development of a semantic ontology and the standardization of ICT communication protocols, facilitating interoperability and the integration of smart solutions across different platforms, permitting the end users to play a pivotal role in optimizing energy efficiency and supporting grid stability.

An exemplary case of digital energy services is the Moneta District in Milan, Italy. This social housing real estate intervention was promoted by REDO SGR SOCIETÀ BENEFIT, with Planet Smart City as the smart solutions partner. In the district of Moneta, thanks to the InterConnect project,

DOI: 10.1201/9788743806851-4

Planet Smart City has integrated advanced digital technologies to enable sustainable living. Over 400 apartments equipped with high-performance insulation and smart monitoring tools allow residents to optimize their energy consumption via a dedicated app. Additional features, including geothermal heating, smart lighting, and innovative irrigation among other services are designed to reduce environmental impact.

The InterConnect project has further enhanced these efforts by introducing AI-managed home devices and also optimizing energy use in common areas and community services. This initiative is not only redefining energy consumption patterns but also creating a collective, environmentally aware lifestyle for the entire community, demonstrating the potential of digital technologies in advancing the future of sustainable urban living.

Keywords: Positive energy districts, energy awareness, digital transition, ancillary services, flexibility, standard adoption, data center design, energy efficiency, carbon footprint computation.

4.1 Introduction

As Europe strides toward its ambitious energy efficiency goals, the integration of digital tools has emerged as a transformative force in reshaping the landscape of energy management. This chapter explores the critical role those innovative technologies – ranging from hardware devices and advanced analytics to Internet of Things (IoT) standardized communication protocols and reference ontologies – play in enhancing energy efficiency across various sectors and untapping the unexpressed potential of flexibility at final stages within the residential sector. With the European Union's (EU) commitment to reducing energy-related CO2 emissions and addressing the rising issue of grid congestion across Europe, digital solutions offer not only a pathway to meeting these targets but also a framework for enhanced collaboration and transparency among stakeholders. By leveraging data-driven insights and automating processes through seamless integration of IoT home, building, and grid devices, we can optimize energy consumption, building a more resilient and sustainable energy future for Europe.

The 20% reduction target set by the EU for 2020 refers to a decrease in greenhouse gas (GHG) emissions, including carbon dioxide (CO2) and other climate-changing gases. This goal drove the information communication technologies (ICT) sector to make significant strides, incorporating modern designs for low-emission and sustainable growth [1].

The ICT sector is the backbone for the digitalization of the energy sector and participates in different ways to this transition. ICT is key for carbon displacements and can support providing energy efficiency (smart grids, smart buildings, intelligent transportation systems, etc.), creating the conditions for a smooth penetration of renewable energy sources, incorporating battery storage, along with electric vehicles and intelligent charging solutions, to reduce energy waste and enhance grid utilization and extending the useful life of transportation and distribution grids.

In this chapter, we will delve into specific case studies and best practices that highlight the intersection of digital innovation and energy efficiency, illustrating how these tools can be effectively leveraged to meet Europe's ambitious energy goals and provide a viable alternative to immediate electricity grid reinforcements.

This report offers a literature review encompassing data available up to October 2024 within the provisional agreements on EMD regulation [2]. However, recognizing the dynamic nature of legislative contexts and the rapid pace of change, the evaluation encompasses the latest legislative amendments, studies, and reports available until the specified date.

4.2 European Energy Policy Framework

The EU has committed to achieving carbon neutrality by 2050, necessitating a reduction in global GHG emissions, particularly from fuel combustion, which accounts for two-thirds of GHG emissions [3]. The energy sector is responsible for more than 75% of the EU's GHG emissions, making a shift toward cleaner energy use crucial for achieving climate goals and reducing dependence on non-EU imports [4].

The residential sector, responsible for a quarter of total energy consumption and GHG emissions, needs swift transformation to achieve climate-neutral goals [5]. The study "EU regions in the transformation towards a climate-neutral future" emphasizes this urgency. Despite fluctuations, households have consistently contributed 25–27% of final energy consumption since 1995, with 27% in 2020 [6].

In September 2023, the International Energy Agency (hereinafter – IEA) presented the Net Zero Roadmap, outlining four key milestones for the global transition to sustainable energy. These milestones include tripling renewable capacity to 11,000 gigawatts by 2030, doubling investment in electricity transmission and distribution grids by 2030, phasing out coal by 2040, and doubling nuclear power from 417 GW in 2022 to 916 GW by 2050 [7].

According to the IEA, the Net Zero Emissions scenario anticipates significant increases in electricity demand due to the electrification of transport and home heating, as well as the widespread adoption of solar photovoltaic (PV) and wind energy. *These factors place added stress on the power grid, and technologies like demand response become crucial to managing these impacts and minimizing the need for expensive new infrastructure.*

The *InterConnect project* [8] (project funded by the Programme Horizon 2020 under Grant Agreement number 857237) aims to pioneer advanced solutions that integrate digital homes and buildings with the electricity sector. It establishes a foundation for future smart energy management by implementing seven large-scale test sites across Portugal, Belgium, Germany, the Netherlands, Italy, Greece, and France. The project's objective is to make efficient energy management accessible to end-users by digitizing homes, buildings, and electric grids through an IoT framework. Leveraging digital technologies like artificial intelligence, blockchain, cloud, and Big Data, based on open standards such as Smart Applications REFerence Ontology (SAREF), ensures interoperability between equipment and systems while maintaining user data privacy and cybersecurity [8].

The sustainability of the InterConnect Project focuses on long-term impacts across various domains: technological, through standardization and regulation bodies; business and market, through go-to-market strategies and policy recommendations; and societal, through awareness and engagement initiatives targeting energy and digital platform end-users [8]. Recognizing the substantial potential of demand-side flexibility to reduce greenhouse gas emissions, this analysis aims to enhance grid efficiency, help solve the grid congestion challenge, reduce investments in areas where consumption can be shifted, and identify and overcome barriers hindering the advancement of demand-side flexibility services across Europe. Prioritizing specific areas is essential for achieving substantial progress in this domain.

According to "Demand response and other distributed energy resources: what barriers are holding them back? 2023 Market Monitoring Report by ACER" multiple European Union (EU), local markets for congestion management services, particularly at the distribution level, are still evolving despite their potential to leverage the flexibility of distributed energy resources [9]. The clean energy package underscores the importance of establishing market-based re-dispatching, known as local markets for congestion management, with only four exceptions outlined in Article 13 of the electricity regulation:

a) no market-based alternative is available;
b) all available market-based resources have been used;
c) the number of available power generating, energy storage, or demand response facilities is too low to ensure effective competition in the area where suitable facilities for the provision of the service are located; or
d) the current grid situation leads to congestion in such a regular and predictable way that market-based re-dispatching would lead to regular strategic bidding which would increase the level of internal congestion, and the Member State concerned either has adopted an action plan to address this congestion or ensures that minimum available capacity for cross-zonal trade is in accordance with Article 16(8) [10].

The European legislative framework establishes key rules and principles for member states, national regulatory authorities, distribution system operators (DSOs), and transmission system operators (TSOs) regarding demand-side response and flexibility through four primary documents:

1. Regulation (EU) 2019/943 of the European Parliament and of the Council of 5 June 2019 on the internal market for electricity (recast) [11] (*hereinafter – the Electricity Regulation*).
2. Directive (EU) 2019/944 of the European Parliament and of the Council of 5 June 2019 on common rules for the internal market for electricity and amending Directive 2012/27/EU (recast) [12] (*hereinafter – the Electricity Directive*).
3. Directive (EU) 2023/1791 of the European Parliament and of the Council of 13 September 2023 on energy efficiency and amending Regulation (EU) 2023/955 (recast) [13] (*hereinafter – EED*).
4. Directive (EU) 2023/2413 of the European Parliament and of the Council of 18 October 2023 amending Directive (EU) 2018/2001, Regulation (EU) 2018/1999, and Directive 98/70/EC as regards the promotion of energy from renewable sources, and repealing Council Directive (EU) 2015/652 [14] (*hereinafter – RED III*).

All four of these legislative measures have been approved and come into force, also representing the EU's broader European Green Deal commitment to achieving climate neutrality by 2050.

Flexibility represents a future-proved digitalization trend for enhancing energy efficiency in the electricity sector.

In recent years, flexibility has emerged as a key aspect of the digitalization trend in energy, particularly valuable for enhancing energy efficiency in the electricity sector, flexibility generally refers to the possibility to adjust the (electrical) consumption or production of an installation or process [15]. Flexibility is key for electricity system operators to maintain a balance between supply and demand, to avoid phenomena such as imbalances, grid frequency deviations, or congestion.

Electrical flexibility can be split into three different categories:

1. Production flexibility: Increase/decrease in the production of power plants (large fossil fuel or nuclear plants but also renewable energy sources (RES) or emergency generators for instance).
2. Consumption flexibility: Also called demand side flexibility (DSF), refers to a temporary increase/decrease in electrical consumption in the industry (e.g., a large industrial process put on hold) or in households (e.g., an electrical boiler starts heating water at a precise point of time).
3. Storage flexibility: Charging/discharging of devices that can store and release energy when needed (pumped hydropower plants, large-scale battery energy storage systems (BESS), domestic batteries, electrical vehicles, flywheels, etc.).

Historically, flexibility was mainly provided by large hydro, fossil fuel, and nuclear power plants. However, with the closure of such type of plants and with an increased penetration of RES (which are less flexible by nature), smaller decentralized assets at the residential level have become an effective alternative for providing flexibility through demand response and emerging as valuable alternatives for stabilizing the grid. These DERs contribute flexibility by dynamically adjusting their output or usage based on grid demand, which can prevent grid overloads and enhance durability during high-demand periods.

When addressing electrical flexibility in the residential sector, it thus mainly refers to consumption and storage flexibility. As this report emphasizes market-based products, *demand-side flexibility* is the main topic addressed and unlocked by digital-driven software solutions.

LCP Delta [16], a prominent research company in the energy sector, employs the outlined schematic below to illustrate general flexibility interactions.

- The *system (or grid) operators (DSO/TSO)* are responsible for distributing or transporting electricity. Sometimes, they face or predict issues on their infrastructure such as grid congestion (for the DSO) or unbalances (for the TSO). In that case, they request flexibility in the market to solve the problem. The *aggregator*[14] contracts with individual distributed energy resources (DER) and electricity demand sites (industrial, commercial, or residential customers) and aggregates them together to operate as a single demand side response (DSR) provider. This way, they can propose flexibility products to the system operators by selling their aggregated flexibility on the market. *Energy Management Systems (EMS)* are the interfaces and intermediary intelligent systems between the aggregators and the aggregated resources. Aggregated resources can be houses for instance but, at the lowest level, they are more likely appliances/devices or processes that can be monitored and controlled. EMS can be *Home Energy Management Systems* (HEMS) in the case of residential customers or Building Energy Management Systems (BEMS) in the case of industrial or commercial sites. Sometimes, a "master-EMS" rules over local EMS to achieve a global optimum.

– The *retailer* (or supplier) is the entity responsible for buying electricity in the wholesale market and selling it to its customers. Retailers often undertake the role of aggregators (if not, they are not directly implicated in flexibility provision).
– The (flexible) *consumer* buys electricity from the retailer and makes use of the EMS to control its devices/appliances or processes. "Flexible" means that the consumer can adapt his behavior by increasing or decreasing energy consumption at certain points in time to offer flexibility to an aggregator.

This diagram below (Figure 4.1) illustrates two different models of flexibility interactions in the electricity sector: LCP Delta and InterConnect. Both the LCP Delta and InterConnect models feature flexibility through aggregators but with different focuses. The LCP Delta model manages interactions between system operators, aggregators, and diverse customers with distributed assets. The InterConnect model emphasizes residential flexibility by using appliances and self-production sources, integrated with Energy Management Systems to respond to grid needs directly.

[14]Other InterConnect deliverables use the term "Flexibility Service Provider (FSP)" to refer to aggregators. As explained in D4.1, both terms are often used interchangeably. Even if both roles are indeed often performed by the same entity, the aggregator sometimes only takes the role of portfolio aggregation and optimization while the FSP act as the market participant that offers the flexibility on the market. The authors of this deliverable decided to use "aggregator" to refer to the entity performing both roles as they find the term to be more self-explanatory.

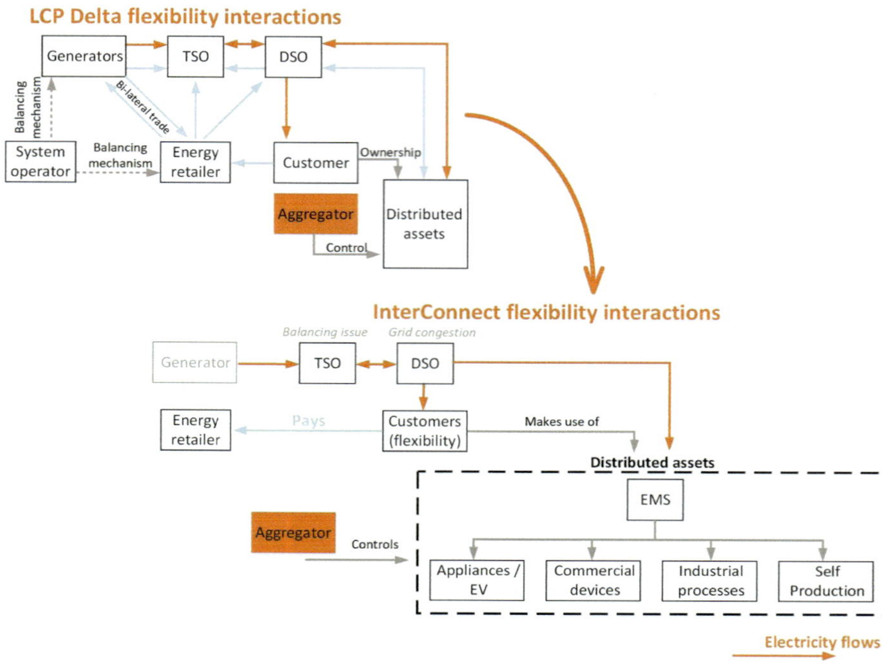

Figure 4.1 Flexibility stakeholders in the InterConnect project.

To foster flexibility within the electricity sector, essential components must be present in the value chain. The process starts with harnessing the potential of *distributed assets*, effectively *managed as a unified entity*. The resulting value is then *traded into a market*.

Utilizing these components as our analytical framework, InterConnect aimed to delineate the boundary between fostering demand side response (DSR) only and the presence of market-based products within the pilots, as experimented also within the Social Housing District of Moneta, Milano, by Planet Smart City and partners.

The diagram below (Figure 4.2) details the components of the value chain of flexibility in the electricity sector.

In a comprehensive flexibility market, the key elements include *customers (1)* possessing monitorable and/or *controllable assets (2)*. These assets are consolidated and *managed as a singular unit (3), optimizing* it according to requirements (4) from entities such as DSOs, TSOs, or others. The optimized *value is subsequently traded* with relevant actors **(5)** like DSOs, TSOs, or others, leading to *compensation for the provided value (6)*.

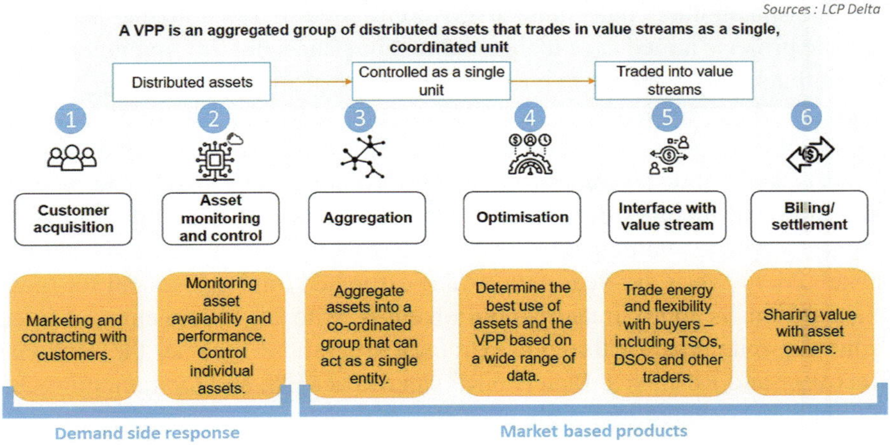

Figure 4.2 Flexibility value chain.

4.3 Milan's Strategy for Carbon Neutrality via Energy Efficiency

In its strategic pursuit of climate leadership, the city of Milan has intensified its efforts to combat climate change by formalizing its commitment to the Climate City Contract on April 22, 2024, Earth Day. This initiative is a key step toward Milan's ambitious objective of achieving climate neutrality by 2030, an acceleration of the goals set out in the city's Air and Climate Plan (PAC), which targets a 2050 completion. The Climate City Contract is not merely a pledge but a framework for action, supported by a consortium of 25 stakeholders, including municipal subsidiaries, universities, urban redevelopment agencies, building cooperatives, and civil society organizations.

Milan's ambitions for achieving carbon neutrality are outlined in the PAC, with key performance indicators (KPIs) [17] focusing on energy efficiency and emissions reduction. Energy efficiency in public buildings: By 2030, the goal is to reduce fossil fuel use by 50% through energy retrofits in public heritage buildings, and over 60,000 square meters of photovoltaic panels will be installed. Urban greening: The Forestami project aims to plant 3 million trees in metropolitan areas, construct green roofs, and expand accessible green spaces. Community engagement: Milan will promote climate awareness, encouraging sustainable lifestyles, circular economy practices, and increased cycling and pedestrian infrastructure to reduce car traffic.

The stakeholders have collectively committed to 157 concrete and measurable projects aimed at reducing carbon emissions and limiting greenhouse gas outputs. These initiatives encompass a range of actions, including "carbon neutral" urban regeneration, modernization of energy distribution networks, the creation of new cycle and pedestrian infrastructures, urban reforestation projects, and public awareness campaigns promoting sustainable energy consumption and mobility behaviors. Of these 157 projects, 51 will be directly overseen by the municipality, while 106 will be implemented by partner stakeholders over a six-year period. The estimated total investment for these interventions is approximately 6 billion euros, with 2.6 billion euros sourced from the municipality through extraordinary funds, such as the National Recovery and Resilience Plan (PNRR) and React EU, and the remaining 3.3 billion euros covered by the engaged stakeholders.

In parallel with these local commitments, Milan has also participated in international initiatives aimed at fostering sustainable urban development. A prominent example is the city's engagement with the global Reinventing Cities competition, an international framework designed to catalyze the regeneration of urban spaces toward decarbonization and resilience. Through this competition, Milan, like other cities worldwide, has identified underutilized urban sites ready for transformation. These areas serve as blank canvases, inviting multi-disciplinary teams – comprising architects, urban planners, developers, investors, environmental experts, creative project leaders, startups, academics, and community associations – to propose innovative models for sustainable urban living. The selected projects are expected to become benchmarks for the future of sustainable cities, demonstrating Milan's commitment to embedding resilience and carbon neutrality into its urban fabric.

These initiatives underscore Milan's multifaceted approach to climate action, combining local efforts with global partnerships to position the city as a leader in the transition toward sustainable urban environments. By participating in both the Climate City Contract and the Reinventing Cities competition, Milan aims to not only achieve carbon neutrality but also act as a catalyst for broader, systemic change in urban development practices through smart energy solutions, digital integration, and inclusive urban planning.

4.3.1 Active citizenship in Milan's energy transition

A cornerstone of Milan's strategy involves empowering citizens as both active consumers and prosumers in positive energy districts (PEDs). Active consumers participate in demand response programs, adjusting consumption based on leveraging smart technologies for energy efficiency. Prosumers go further by generating their own energy, typically through PV installations, and contributing to energy-sharing networks that enhance grid resilience and social cohesion. This shift in consumer behavior reflects insights from the EU-funded InterConnect project (Grant No. 857237), which emphasizes the need for policy support in decentralized energy markets and highlights successful and unsuccessful strategies for active consumer engagement. For additional context on these decentralized marketplaces, see Section 4.5.

4.3.2 Role of REDO SGR in sustainable urban development

A key player in Milan's Climate City Contract is REDO SGR, a leading developer focused on affordable and sustainable housing. REDO's projects integrate energy-efficient designs, innovative construction materials, and advanced technologies to support Milan's carbon neutrality goals while addressing urban housing needs. Through energy-monitoring systems and renewable energy production, REDO enables residents to actively manage their energy usage, contributing to the city's sustainability objectives. This approach aligns with broader goals to establish positive energy districts, fostering social inclusion and environmental stewardship within Milan's evolving urban landscape.

4.4 Innovative Urban Regeneration and Digital Integration in Milan's Transition to Sustainable, Resilient Communities

4.4.1 Towards the digitization of services: REDO SGR and Planet Smart City collaboration

Since 2017, REDO has collaborated with Planet Smart City, a Turin-based competence center renowned for its expertise in creating smart urban districts. This partnership has fostered the development of a forward-thinking approach to urban planning, emphasizing the integration of technology and sustainability. One of the key outcomes of this collaboration was the successful joint participation in the global "Reinventing Cities" competition, which promotes the transformation of under-utilized urban sites into sustainable and resilient landmarks.

Together, REDO and Planet Smart City, as part of a large consortium with over 50 partners, have achieved prominent awards in the competition, most notably for projects like:

- *L'Innesto* in the Greco-Breda area is a community-centered project that prioritizes carbon neutrality.
- *Green Between* in the Crescenzago district integrates expansive green spaces and energy-efficient infrastructure.

Through these projects, REDO has demonstrated its capacity to lead in the domain of affordable and sustainable urban development, leveraging its expertise to deliver high-quality housing solutions that align with Milan's environmental and social goals. This continued commitment to innovation underscores REDO's pivotal role in shaping the future of urban living, where affordability, sustainability, and technology converge to create resilient and inclusive communities.

4.4.2 Towards positive energy districts: ARIA Project for the Reinventing Cities competition

The ARIA project stands out as Milan's first carbon-negative urban development, setting a new benchmark for sustainable construction in the city. This achievement will be made possible through the implementation of the Ectogrid energy district and photovoltaic systems powered by Renewable Energy Communities, which will generate 100% renewable energy, including 30,000 square meters of rooftop solar panels. The initiative is set to obtain LEED Gold certification for its buildings and the LEED for Cities and Communities certification for its master plan, demonstrating its comprehensive commitment to sustainability. A key aspect of ARIA involves integrating green infrastructure, including parks, vegetable gardens, green facades, and dedicated spaces like the "Air Factory" to purify indoor air by filtering atmospheric contaminants. The project will provide over 1,000 affordable housing units, with rental prices for 60 sqm apartments set at approximately âĆň500 per month and sale prices at âĆň2,500 per square meter. In addition to 50,000 sqm of pedestrian-friendly spaces and 30,000 sqm of green areas, the development will host 3,000 new residents and create 30,000 jobs. Its carbon footprint is projected to be 52% lower compared to traditional construction projects, with at least 30% of construction materials coming from recycled sources and half of landfill waste redirected to recycling. Moreover, the reuse of groundwater from the Borgognone River will decrease water consumption per resident by 30%.

This ambitious initiative sets the stage for further innovative projects, with the same mission of REDO to promote the adoption of digital services in other districts such as Moneta, Merezzate, and Quid Quintiliano.

4.4.3 Moneta case study: A model for sustainable and connected living

The Moneta District project has been developed by REDO SGR as the project leader, with Fondazione Housing Sociale serving as the Technical and Social Advisor. The design phase is being managed by architectural firms Mpartner, Beretta Associati, and MAB Arquitectura, while Planet Smart City has contributed its expertise as a smart technology consultant. This collaboration brings together diverse skills to create a sustainable and technologically advanced urban environment.

In the Moneta district, technology plays a pivotal role in enhancing sustainability and improving residents' quality of life. The development features 400 apartments, all meeting Class A energy efficiency standards, and equipped with high-performance thermal and acoustic insulation. Technological solutions focus on three main areas:

1. Environmental impact: The district utilizes a geothermal heat pump, underfloor heating, smart LED lighting for public areas, shading systems, and an innovative irrigation system, all designed to minimize environmental impact.
2. Consumption monitoring: The Redo App is a key tool for residents, allowing them to track and customize their energy and water consumption, reducing waste and cutting costs.
3. Security and community: Common areas are equipped with free Wi-Fi and video surveillance systems to ensure safety. The Redo App also enables efficient management of communal spaces and provides information about neighborhood events, facilitating community engagement and social activities among residents.

These technological and design choices aim to create a model of sustainable living that aligns with broader goals of environmental and consumption responsibility and social cohesion.

4.4.4 Planet App in Moneta district

In the Moneta Social Housing District, the Planet App has been a crucial technological tool, enhancing residents' smart living experience even before the InterConnect project launched its experimentation phase in the area. The App allows residents to interact seamlessly with their environment, providing a comprehensive suite of features designed to improve both sustainability and convenience.

Through the Planet App, residents can monitor air quality, control home devices, track energy consumption, and make reservations for communal spaces. It also serves as a communication hub, allowing residents to stay informed about neighborhood events, share updates with neighbors, and alert emergency contacts if needed.

Key district amenities include free Wi-Fi, a video surveillance system, electronic keys for common areas, and high-speed fiber optic internet. This next-generation broadband connection enables fast and reliable internet access, making it easy for residents to stream content and remote management of smart home functions.

A notable feature, called Flexi, was introduced thanks to the InterConnect project. It allows energy flexibility within the district by equipping each apartment with advanced washing machines connected to an algorithm driven by machine learning, which optimizes washing cycles by adjusting them to energy-efficient time slots – periods, when energy demand on the grid is lower and renewable energy supply is higher. This reduces energy consumption and supports the district's sustainability goals by aligning power usage with grid signals.

In the electricity sector, newly connected decentralized energy sources can be effectively managed through a layer of digital communication infrastructure. This also applies to consumer and industrial assets, whether it is a large industrial load, a smart home device, or an electric vehicle charger.

In the electricity sector, the district's digital communication infrastructure enables the effective management of decentralized energy sources, including consumer assets like smart home devices and EV chargers, alongside larger industrial loads. As digitalization permeates every aspect of the energy supply chain, vast data streams from connected assets can be analyzed, and leveraged to enhance grid flexibility and efficiency. This could ultimately reduce the need for extensive grid upgrades, facilitating the integration of renewable energy sources and the electrification of sectors like transportation. To facilitate seamless information exchange throughout the energy system, interoperability must be approached from multiple dimensions. This includes the interoperability of physical devices, the data formats and information models employed, and the overall harmonization of business processes and regulatory frameworks. The InterConnect project has enhanced this approach using the Smart Grid Architecture Model (SGAM) as a guiding framework.

It is crucial to recognize that various dimensions of interoperability must work together harmoniously. Therefore, analyzing the barriers to adopting interoperable systems in the energy sector requires a comprehensive consideration of technological, economic, and regulatory factors.

This system-level interaction was enabled by a Semantic Interoperability framework that defined a data model and type for all players of the electrical value chain, from the production up to the final energy consumption units, such as home devices (EV charging stations, electrical heat pumps, with goods in general).

4.5 Interconnect Experimentation at Moneta District: The Flexi Feature

The InterConnect project (EU-funded under Grant Agreement number 857237) provides valuable insights in its D9.5 Policy Recommendations for fostering decentralized energy marketplaces sharing the analysis of a successful pilot project in the residential sector reveals that the lack of an aggregator or automation solution is a significant barrier to activating demand-side flexibility services, contributing to project failure. For these services to thrive with active consumer participation, both DSOs and aggregators must emphasize direct and consistent communication with consumers, utilizing user-friendly digital tools and regular notifications. In one project, the absence of automation required consumers to manually activate flexibility within 15 minutes of notification, a highly impractical expectation that undermined the pilot's success. Conversely, the Planet App focused on gamified automation, promoting inclusiveness and enhancing the experiment's effectiveness.

Within the InterConnect project, Planet Smart City launched and tested in real-life environment a digital service to unlock energy flexibility for the residents of the district, after seamless integration of IoT dispatchable electrical loads controlled via the Planet App, following the CEN EN 50631 standard protocol and SAFER4ENER reference ontology.

Residents engaged in a gamified experience through the Planet App, receiving push notifications and content-driven insights to increase awareness. Missions were set up to encourage residents to define which devices they wanted to let available for Demand Response mechanisms. People could program one day ahead of the running of washing cycles for a dishwasher or a washing machine. Data-driven machine learning algorithms allocated them at optimal times to optimize energy consumption across the entire complex, responding to some synthetic grid congestion signals. In addition, residents were also...

Residents were also informed about the carbon intensity of the distributed electricity for the upcoming 24 hours and were encouraged to adopt sustainable behaviors to contribute to peak shaving. (carbon intensity or carbon footprint is more accurate here as it talks about the amount of CO2 emission associated with electricity supply).

This service defines a new lifestyle and consumption pattern for an entire community, acting collectively with attention to environmental and economic savings, orchestrated by Planet Smart City acting as a balance service provider mimicking digital services to enhance grid stability.

The project was launched in Moneta in October 2021 and extended to Merezzate and Quintiliano in the autumn of 2023, in collaboration with Planet Smart City, REDO'S partner for smart solutions, Whirlpool and Miele as the equipment provider, and Research on the Energy System (RSE), a company of the GSE (Gestore Servizi Energetici) Group that conducts research in the electro-energy sector.

4.5.1 The model of the Citizen's Lab

The Citizen Labs format is an initiative designed by the community development team at Planet Smart City. Its main objective is to provide participants with tools and knowledge on specific topics that are often complex to understand.

The program includes regular meetings led by experts or project partners who share their expertise to make technical and scientific subjects accessible while promoting responsible lifestyles.

In addition to the educational goal, the format also aims to foster community engagement and activate the social capital present in each community through group activities and bottom-up planning managed by community managers.

For the InterConnect project, the citizen lab was linked to the theme of energy. In the Moneta social housing district we embarked on this journey and, for over 24 months, Planet Smart City accompanied step by step the wider community of residents and those who have chosen to join the project activities to deep dive into energy.

The Citizen Lab are fixed-term events which have the aim of accompanying a community along a path that aims above all to generate relationships, knowledge, and awareness.

Relationship, knowledge, and awareness are three keywords for the format led by the community development business unit of Planet Smart City which proposes and plans these events. Over time the Citizen Lab format has involved around 50 inhabitants of the Moneta district. Team building activities aimed at strengthening relationships between the inhabitants themselves, training activities on specific contents, sometimes complex but fundamental for the InterConnect project, and to do this we also made use of the collaboration of experts among our partners such as RSE.

However, we also had fun with the group of experimenters as we like to define those who have chosen to join the InterConnect project by involving them in playful and gamification activities always oriented toward developing and generating awareness on specific themes.

These labs are often used in projects focused on technological themes, with the intent of ensuring a full understanding of the proposed initiatives, maintaining ongoing engagement, and creating a space for collecting data and feedback from participants.

4.6 Conclusion, Results, and Openings for Future Research Projects

Promoting interoperability, standardization, and open market opportunities in the electricity sector are essential for unlocking the full potential of digitalization, thus tackling grid retrofitting urgency by optimizing the existing infrastructure. As the industry increasingly depends on data exchange among various systems and stakeholders, the demand for interoperable solutions intensifies. However, achieving this goal necessitates coordinated efforts across technology, policy, and market dynamics.

The complexity of the regulatory landscape, along with diverse business interests and technical obstacles, highlights the importance of a governance framework that encourages interoperability together with the adoption of standardized and digitalized means by every player of the energy value chain, from production up to the final users. This involves creating incentives for both regulated and unregulated entities, developing specific standards for distinct use cases, and establishing minimum interoperability requirements. Additionally, it is critical to address cybersecurity and privacy issues to ensure secure and reliable data exchanges.

Furthermore, effective governance mechanisms must be put in place to monitor and update standards, ensuring they stay relevant and foster innovation. Labelling and certification systems can also facilitate the adoption of interoperable solutions by minimizing information asymmetry and enhancing transparency for all stakeholders.

The analyzed projects, particularly Milan's ambition, aim to integrate energy efficiency throughout the entire life cycle by involving different stakeholders in the energy value chain. Achieving this requires creativity, openness, and active, transparent communication that delivers accurate and diverse added value. It is important to ensure open communication that provides accurate and diverse added value not only for household representatives but for all market participants, whether regulated or not. While ICT is a vital tool, transparent and effective governance is a necessary platform to ensure that everyone can participate.

By focusing on these key principles and fostering collaboration across all relevant parties, market-driven opportunities can propel the electricity sector toward a more interconnected, efficient, and resilient future. The shift to a digitalized, interoperable grid not only promises to enhance grid reliability but also creates avenues for innovation, cost savings, and improved sustainability outcomes.

Acknowledgments

Authors would like to express their gratitude to all esteemed colleagues, both internal and external to Planet Smart City, whose contributions were instrumental in the completion of this chapter on Digital Services for Energy Transition and wish to acknowledge the invaluable guidance and encouragement provided by David Rua, as well as the expertise of all members of the InterConnect community. Their extensive knowledge of energy systems and digital technologies has profoundly influenced the direction of this chapter. I am particularly grateful to colleagues from RSE (Renato Urban, Paolo Serafini) for their ongoing and constructive dialogue on energy systems.

This chapter would not have been realized without the unwavering support of Planet Smart City's expertise and commitment to operational excellence, extending our deepest thanks to those involved in community development (Giulia Stefanelli, Jacopo Rossi), research (Barbara Melis, Simone Panfilo), energy services (Adriano Bergese), and the digital uptake of innovative services (Antonio Pio Roseti, Cristina Ghione, Gabriella Piccolo). Their contributions were essential in enabling the implementation of digital services within the energy domain, which formed the foundation for the analysis and case studies presented in this chapter, based on data from real consumers.

Special thanks to REDO SGR (Andrea Vecci, Victor Avila, Paola Amadeo, Olimpia DeAmbrosis) for their endorsement of Planet Smart City's efforts to bring these advanced digital energy features and systems into practical application. Their support has been pivotal in fostering the digital transition within the energy sector, highlighting the critical role of end users within the broader value chain.

Finally, we would like to convey my sincere appreciation to the end users in the residential districts of Moneta, Merezzate, and Urbana (Milano) for their active participation in utilizing the digital services provided. Their constructive feedback and valuable observations have been essential in enhancing the digital domain and ensuring its continued development.

References

[1] International Telecommunication Union (ITU), 'Report on Climate Change', Oct. 2008.

[2] European Parliament and Council, 'Proposal for a Regulation amending Regulations (EU) 2019/943 and (EU) 2019/942 as well as Directives (EU) 2018/2001 and (EU) 2019/944 to improve the Union's electricity market design - Analysis of the final compromise text with a view to agreement', available at: https://data.consilium.europa.eu/doc/document/ST-16964-2023-INIT/en/pdf. [accessed Feb. 2, 2024].

[3] European Parliament News, 'Reducing carbon emissions: EU targets and policies', updated: Oct. 27, 2023, available at: https://www.europarl.europa.eu/news/en/headlines/society/20180305STO99003/reducing-carbon-emissions-eu-targets-and-policies [accessed Jan. 14, 2024].

[4] European Commission, 'Renewable Energy Targets', available at: https://energy.ec.europa.eu/topics/renewable-energy/renewable-energy-directive-targets-and-rules/renewable-energy-targets_en.

[5] Climate Policy Info Hub, 'Household Contribution to Buildings' Carbon Footprint', available at: https://climatepolicyinfohub.eu/household-contribution-buildings-carbon-footprint.html [accessed Oct. 14, 2024].

[6] Eurostat Statistics Explained, 'Energy Consumption in Households', data extracted in June 2023, available at: https://ec.europa.eu/eurostat/statistics-explained/index.php?title=Energy_consumption_in_households[accessedOct.20,2024].

[7] International Energy Agency, 'The path to limiting global warming to 1.5 °C has narrowed, but clean energy growth is keeping it open', Sep. 26, 2023, available at: https://www.iea.org/news/the-path-to-limiting-global-warming-to-1-5-c-has-narrowed-but-clean-energy-growth-is-keeping-it-open [accessed Jan. 21, 2024].

[8] InterConnect, available at: https://interconnectproject.eu/about/#results [accessed Feb. 10, 2024].
[9] ACER, 'Demand response and other distributed energy resources: what barriers are holding them back? 2023 Market Monitoring Report', Dec. 19, 2023, available at: https://www.acer.europa.eu/news-and-events/news/acer-offers-do-list-remove-barriers-hinder-demand-response-new-entrants-and-small-players [accessed Oct. 23, 2024].
[10] Regulation (EU) 2019/943 of the European Parliament and of the Council of June 5, 2019, 'On the internal market for electricity', available at: https://eur-lex.europa.eu/legal-content/EN/TXT/PDF/?uri=CELEX:32019R0943 [accessed Jan. 14, 2024].
[11] Regulation (EU) 2019/943 of the European Parliament and of the Council of June 5, 2019, 'On the internal market for electricity (recast)', available at: https://eur-lex.europa.eu/eli/reg/2019/943/oj [accessed Feb. 2, 2024].
[12] Directive (EU) 2019/944 of the European Parliament and of the Council of June 5, 2019, 'On common rules for the internal market for electricity and amending Directive 2012/27/EU (recast)', available at: https://eur-lex.europa.eu/legal-content/EN/TXT/?uri=celex%3A32019L0944 [accessed Feb. 2, 2024].
[13] Directive (EU) 2023/1791 of the European Parliament and of the Council of Sep. 13, 2023, 'On energy efficiency and amending Regulation (EU) 2023/955 (recast)', available at: https://eur-lex.europa.eu/legal-content/EN/TXT/?uri=OJ%3AJOL_2023_231_R_0001&qid=1695186598766 [accessed Feb. 2, 2024].
[14] Directive (EU) 2023/2413 of the European Parliament and of the Council of Oct. 18, 2023, 'Amending Directive (EU) 2018/2001, Regulation (EU) 2018/1999, and Directive 98/70/EC as regards the promotion of energy from renewable sources, and repealing Council Directive (EU) 2015/652', available at: https://eur-lex.europa.eu/legal-content/EN/TXT/?uri=OJ%3AJOL_2023_231_R_0001&qid=1695186598766 [accessed Feb. 2, 2024].
[15] Next Kraftwerke, 'Flexibility in the electricity sector', available at: https://www.next-kraftwerke.be/knowledge-hub/flexibility-electricity-sector.
[16] Delta-EE, available at: https://delta.lcp.com/.
[17] Climate-KIC, 'A new air and climate plan for Milan', available at: https://www.climate-kic.org/news/a-new-air-and-climate-plan-for-milan/.

5

The General Theory of Automation Systems with Energy- and Resource-Saving Functions

I. Nevliudov[1], Y. Romashov[1], and T. Tkachova[1,2]

[1]Kharkiv National University of Radio Electronics, Ukraine
[2]XLIM Research Institute, France
E-mail: igor.nevliudov@nure.ua; yurii.romashov@nure.ua; tkachova.tetiana@gmail.com

Abstract

The principal idea of this research is to develop the theoretical foundations for the development and implementation of automation systems with energy and resource-saving functions for industrial and house-holding applications, including supporting different intelligent environments such as smart cities and energy-saving buildings. These energy- and resource-saving functions are achieved on different operational modes through the correspondent optimal choices of the control programs and automatic governors' parameters. The development of this principal idea is through performing the general theory for the development of automation systems with additional energy and resource-saving functions on the basis of optimal control problems for transient and stationary processes. The approaches to consider the optimal control problems are developed based on reducing these problems to the resolving of the related constrained optimization problem suitable for numerical methods usages. Providing energy-saving functions in electric drives with direct current electric motors and resource-saving functions in water level controllers, which are widely used in different intelligent environments such as smart cities and energy-efficient buildings are considered as the possible applications of the developed general theory and approaches. The possible effects due to the usage of the proposed theory and approaches are shown for

DOI: 10.1201/9788743806851-5

automated systems using computer simulations, which are conducted using Scilab free open code software for scientific and engineering computing.

Keywords: Automation, energy efficiency, resource saving, mathematical modelling, optimization, optimal control, automatic governors.

5.1 Introduction

Energy and resource efficiencies are at the core of the recent decade's ideas about providing sustainable development, and such ideas are considered at present by progressive societies as the modern trends leading to the related targets near future, as it is discussed for different activities' fields like in the articles [1, 2] and others. Taking into account this circumstance, all research on providing energy or (and) resource efficiencies is relevant today, and all of it is relevant in any case due to promotion for increasing general efficiency both for businesses in different fields and both for households. Thus, this research about providing energy and resource efficiencies through the development of suitable automation systems is relevant due to agreement with the modern general trends of progressive societies, targeted to sustainable development ideas.

5.1.1 Research backgrounds and literature reviews

It is clear that energy and resource efficiencies of objects are the results of the inherent properties of these objects, so such efficiencies must be archived through the appropriate design of the objects. Although energy and resource efficiencies have always been of interest, only sustainable development ideas formed in recent decades have placed these efficiencies at the core of all activities in different fields, including also processes of designing, manufacturing, and exploiting engineering objects. The levels of energy and resource efficiencies of the objects are significantly predefined by the used designing principles, and it is necessary to develop the relevant designing principles putting such efficiencies as one of the main purposes of the design to have the engineering object agreed with the sustainable development ideas. On the other side, a lot of operations of present engineering objects were designed and manufactured a long time ago without energy and resource efficiencies as the main principles of the design, and the renovation of such objects is a separate problem in providing the sustainable development of the societies at present.

Buildings are interesting examples of engineering objects operating for a long time after designing and manufacturing, but also widely designed and manufactured at present. A lot of different designing and manufacturing principles are developed and implemented at present to provide energy efficiency for new buildings, and the article [3] is one illustration of it. At the same time, a lot of old buildings are in exploitation at present, so the renovation of a building to provide energy and resource efficiencies is another more difficult problem, which is solved at present through the modernization of automation systems [4]. Modernizing automation systems is the most suitable way for building renovation to provide higher energy and resource efficiencies, because such a way does not require the remanufacturing of the building structures, and it can be applied to any existing buildings independently of their age. Besides, it is possible to implement modernizations of automation systems periodically during the time of building operation to increase energy and resource efficiencies taking into account recent achievements in automation engineering. Improving automation systems allows for achieving higher energy and resource efficiency of buildings through efficient lighting systems, water conservation and efficiency measures, and implementation of energy management [4]. Opportunities for automation in achieving energy and resource efficiencies in the building led naturally to smart buildings, which are initially designed and constructed with improved embedded computer-integrated automation systems to provide intelligent management [4]. The possibilities of energy and resource efficiencies of the buildings are improved quickly due to the development of automation engineering, so there is a lot of research in these fields at present. The research [5] shows the trends in the implementation of building energy and climate automation control systems for improving energy efficiency. It is natural that the purpose of the building influences the design of automation systems to achieve the maximum possible energy and resource efficiencies. So, in the research [6], it critically evaluates the accuracy of energy savings estimation due to building automation and control systems implementations for office buildings; the particularities of implementation of some novel smart control strategies in a hospital-integrated energy system are discussed in the research [7]. Article [8] shows the development of automation engineering through computer-integrated and Internet of Things technologies implementations for smart building control. Buildings are naturally envisaged for people, which can have influences on building automation systems for energy and resource efficiency, so building automation has social aspects, which are discussed in the articles [9, 10].

Fields about water handling give clear examples of resource efficiency demands at present in agreement with the Sustainable Development Goals [11, 12]. Water efficiency is important for the buildings [4] at least to minimize the exploitation costs, but it covers only the water handling just inside the separate consumer, although buildings can be very large and can have a lot of sub-consumers. The novel automation engineering technologies, including the computer-integrated and the Internet of Things, allow optimal water management, as discussed in the article [13]. The water supply optimization can involve also the energy efficiency criteria [14], so the energy and water resource efficiencies are closed and complement each other in a lot of applications, and the relevant automation and control strategies are the core of optimal water usage.

Smart environments are built not only in the urban areas, but also in the rural areas and agriculture, and the related automation systems are at the core of them, as highlighted in the article [15]. Agricultural production is based on the usage of different machines with electric motors, and to provide energy efficiency it is necessary to design and implement the relevant electric drives, as well as energy management for different operations, as it is shown in the article [16] on the particular example about the manufacturing of animal feed. Water efficiency is significantly important for agricultural activities due to the necessity of large water amounts, and a lot of existing researches show that implementation of improved controllers into the related automation systems allows for providing energy and water efficiency of agriculture, including for irrigate systems, as discussed in the articles [17, 18].

Different kinds of vehicles represent interesting examples of objects requiring energy efficiency to improve their important exploitation characteristics, and implementation of relevant automation systems is one of the ways to provide this energy efficiency, as is shown in a lot of existing publications. In the research [19], it is highlighted that vehicle automation allows providing energy saving for electromobility, including through decreasing energy consumption for additional vehicle functions such as heating and air conditioning, and modeling of the energy consumption of heating, air conditioning, and vehicle automation is considered as an extension of electric vehicles inherent abilities. From the research [20] it is seen that energy consumption can be considered as the core of a novel ecological driving strategy for electric vehicles for urban traffic conditions, including inherent for smart cities. It is natural that the energy-saving properties of electric vehicles can be achieved through the relevantly developed suitable automation systems, and the optimization for motor speed controllers is required to have it, as discussed in the research [21]. Of course, the energy management strategies for electric vehicles must be directed not only to the high fuel efficiency but to the lifetimes of the fuel cell system and the batteries to provide resource efficiency, as discussed in the research [22]. It is clear that the automation of vehicles significantly depends on the energy source type of rotation couples, so a lot of researches consider particular problems for different kinds of vehicles, like the article [23], which deals with the optimal control approach to design energy management strategies for hybrid electric vehicles.

The unmanned technical objects principally require full automation, but last time energy and resource efficiencies are in the attention as the complementary functions improving the exploitation characteristics, as discussed in the article [24]. Of course, the human-machine interactions related to the automation systems are the most important for unmanned technical objects [25].

The effects of automation systems implementations are usually expected as the exclusion of people from the monotonous or (and) dangerous processes, but last time, energy and resource efficiencies are considered as principal expectations from automation systems. So, energy and resource efficiency criteria have priority attention last time in research about optimal control. For example, indoor temperature control is considered in the research [26] relevant to energy efficiency, to minimize energy consumption during the operation. A similar situation is in the article [27], but energy efficiency is considered the principal core of climate control systems. The energy efficiency of the production is also considered the principal purpose for multi-criteria optimization of the pneumatic control for compressed air systems in the research [28]. It is difficult to give common approaches providing energy and resource efficiencies for technical objects with different energy sources, so a lot of researchers consider particular problems like the research [29], which deals with the energy-saving control for hydraulic servo-axes.

Electrical drives are the most used for various industrial and household purposes in different technical systems, so the energy efficiency of electromechanical systems is often considered a research subject at present. For example, it is natural that energy efficiency controls are considered often for electric drives, like in the research [30]. The energy and resource efficiencies are mutually dependent on the electromechanical systems because the energy efficiency leads to battery charging cycles decreasing and as a result – to increasing battery life, as discussed in the research [31] as an example.

Mathematical models representing the automation objects can be nonlinear and they can have other properties producing difficulties of analysis, so the analytical approaches have the limitations for considering optimal control problems. Taking into account this circumstance, the numerical methods are more suitable to provide uniform approaches for the optimal control problem solutions, as discussed in the research [32, 33] for example. The research [34] and their continuation [35] demonstrate the possibilities of the development of universal approaches based on different numerical method combinations to build the optimal control for transient processes in objects.

5.1.2 The research purpose, methods, and tasks

It is understandable that the relevant approaches must be used to design automation systems with inherent properties of energy and resource efficiencies, and such approaches must be based on the corresponding general theory. The traditional well-known all-conventional general theory of automated systems is based on the stability theory like in the research [36] for an example, and on the optimal control theory like in the research [37] for an example. The stability theory allows designing automated systems to stabilize the controlled parameters of the objects during operation, but these automated systems have no energy and resource efficiencies naturally so to provide it, the relevant development of the existing general theory of automation systems must be accomplished. The optimal control theory allows to design the controls providing the transient modes of objects relevant to the considered optimization criteria, but most of the results are for the linear systems predominantly, and the numerical methods are not fully developed to consider the complicated, including nonlinear systems. Taking into account all these circumstances, the principal purpose of this research is to develop the general theory of automation systems providing energy and resource efficiency functions for different technical objects, for industrial and house-holding applications, including those related to supporting different intelligent environments such as smart cities, smart farming, energy-saving buildings, Internet of Things automation, and others. While developing such general theory, i1t is assumed that automation objects' mathematical models will be represented using methods related to the ordinary differential equations, to have energy and resource efficiency automation through the consideration of the related optimal control problems reduced to the finite-dimension optimization. It is assumed also to use only numerical methods to provide the computer-aided automation engineering analysis for the designing of automation systems with energy efficiency and resource-saving properties for different technical objects under industrial scales.

To achieve the aforementioned purpose of this research, the following tasks are envisaged:

- Develop the general theory for the mathematical modeling of automation objects and automatic governors, introduce the notions about transient and steady-state operating modes;
- Consider energy and resource-saving optimal controls for transient modes of automation objects by means the reducing to the resolving constrained finite-dimensional optimization problem suitable to use the related numerical methods;
- Offer the parameters choosing approaches allowing to have the suitable automatic governors providing energy- and resource-saving on the steady-state modes of automation objects by means the reducing to the equivalent resolving constrained finite-dimensional optimization problem suitable to use the related numerical methods;
- Illustrate the possibilities of energy-saving optimal controls for the angular velocity of the electric drive with direct current motor under the transient modes, as the particular examples of the applications of the developed general theory;
- Show the opportunities of the water consumption decreasing due to the relevant choosing of the parameters of PI-governor for automated water level control in the vessels, as the particular example of the developed general theory application to provide the resource saving.

These research tasks cover all the principal theoretical and practical aspects of the developed general theory of automation systems with inherent energy and resource-saving properties. Besides, the considered examples of the energy-saving automated electric drive have their own value due to wide opportunities for implementations in different technical objects for industrial and household purposes. The example of water level automated control systems is related to the decrease in water consumption, and it has also own value due to limited water resources. Thus, all the envisaged examples of the developed general theory of automation systems with intrinsic energy and resource savings align with the Sustainable Development Goals [38].

5.2 Mathematical Modelling of Automation Objects

The development of automation systems with inherent energy- and resource-saving functions requires the relevant consideration of the properties of automation objects through the appropriate mathematical modeling of the intrinsic processes in these automation objects. The all-conventional view on mathematical modeling is through the nature of the considered processes – mechanical, thermodynamic, chemical, and others. At the same time, the automation objects usually involve reciprocal processes with different natures, so it is difficult to propose a universal methodology for the mathematical modeling of automation objects on the basis of traditional views. So, the general view on the mathematical modeling of automation objects is proposed as the principally required constituent to substantiate the automation systems with energy and resource-saving functions.

5.2.1 Processes in automation objects

Mathematical modeling is envisaged generally to represent the processes to research their properties by means of the related analysis. So, the mathematical modeling of the automation objects is reduced to the mathematical modeling of their intrinsic processes. Because processes with different natures can be inherent for automation objects, it is difficult to propose universal approaches for mathematical modeling of them. Taking into account this circumstance, the principal proposed idea is to divide the mathematical modeling of the processes inherent in the researched automation object from the mathematical modeling of the automation object as its own essence to represent their intrinsic properties to design a suitable automation system with energy and resource-saving functions. Thus, peculiarities of the mathematical modeling for processes with different natures are not considered here, because it is in mathematical physics and other related fields, but the general approaches are considered how to lead from the mathematical models of processes to the mathematical models of automation objects.

The processes in automation objects are the same as the processes in other objects, so it is possible to use the traditional way to define the processes by means of the set of state parameters depending on the time that is to say as the set of the functions of the time:

$$\mathbf{q} = \mathbf{q}\left(t\right), \tag{5.2.1}$$

where t is the time; $\mathbf{q}$ and $\mathbf{q}\left(t\right)$ are the vector of state parameters involving the state parameters of the process inherent for the considered automation object and the related vector function representing the dependencies of these state parameters on the time.

The reference of the time t to represent the evolution of the vector of state parameters (5.2.1) is arbitrary, and it can be chosen in agreement with the peculiarities of the researched operation mode of the considered automation object. The senses and the count of the state parameters involved in the vector (5.2.1) are defined by the nature of the processes inherent in the considered particular automation object. The definition (5.2.1) is general, and it is applicable to any processes possible for the automation objects.

The mathematical model of the processes must allow for defining the state parameters (5.2.1), and the consideration of the corresponding system of ordinary differential equations is the most suitable way to do it. It is principally that the system of ordinary differential equations defining the state parameters (5.2.1) of the considered processes must have a special view of the processes considered as constituents of automation objects. Indeed, the state of any automation object is supported by means of the specially formed controlling influences, so for any automation object, it is required to introduce the controlling parameters, which are the set of independent parameters defining such controlling influences, as it is discussed in the researches [34, 35]. It is assumed in this research that the controlling influences of the automation objects can be defined by means of the related vector:

$$\mathbf{u} = \mathbf{u}\left(t\right), \tag{5.2.2}$$

where $\mathbf{u}$ and $\mathbf{u}\left(t\right)$ are the vector of controlling parameters of the considered automation object and the related vector function representing the dependencies of these parameters on the time.

The senses and the count of the controlling parameters (5.2.2) are defined by the nature of the processes inherent for the considered particular automation object, as well as by the structure of the related automation system. Nevertheless, it is principally important to define the vector of controlling parameters (5.2.2) to construct the mathematical model of the automation object. Exactly the involvement of the vector of controlling parameters (5.2.2) in the differential equations, defining the vector of state parameters (5.2.1), allows to consider these equations as the mathematical model of the processes in some automation object. So, the most general representation for mathematical models of the processes inherent in the considered automation objects can be through the differential equations having the following view:

$$\frac{d\mathbf{q}}{dt} = \dot{\mathbf{q}}\left(t, \mathbf{q}; \mathbf{u}\left(t\right)\right), \quad \mathbf{q}\left(t_0\right) = \mathbf{q}_0, \tag{5.2.3}$$

where $\dot{\mathbf{q}}\left(t, \mathbf{q}; \mathbf{u}\left(t\right)\right)$ is the given vector function defining the velocities of changes of the state of the processes depending on the time, the current state, and the controlling influences in the considered automation object; t_0 and

$\mathbf{q}_0$ are the given initial time and the given related vector representing the initial state of the considered automation object assumed to consider their researched operation mode.

The view of the differential equations (5.2.3) is defined by the nature of the inherent processes as well as by accepted simplifications and schematizations accompanying any mathematical modeling, and the particularities of these differential equations (5.2.3) are related to each of the considered automation objects. To build the differential equations (5.2.3), it requires the usage of mathematical physics, such as the theory of heat transfer, the theory of elasticity, the theory of electromagnetic fields, chemical kinetics, the physics of nuclear reactors, or others related to the processes inherent for the considered automation object. At the same time, the particularities of the differential equation (5.2.3) are not important here, but it is important that the mathematical model (5.2.3) allows us to predict the state (5.2.1) of the processes in the considered automation object for any given vector of controlling parameters (5.2.2).

5.2.2 Automation objects modeling

It is necessary to highlight that any automation object envisages not only the controlling parameters but also the parameters that must be controlled during the operation. At the same time, these controlled parameters are not involved in the mathematical model (5.2.3), and it allows representing only the processes within the automation object. Therefore, the mathematical model (5.2.3) does not represent the mathematical model of the automation object itself. Nevertheless, the properties of any automation object are defined by the inherent processes, so on the basis of the previous research [34, 35], it proposed the generalized approach for the mathematical modeling of automation objects by means of a supplement of the mathematical model (5.2.3), representing the processes inherent for this automation object.

As noted above, for any automation object, it is necessary to envisage the parameters that must be controlled during the operation, and these parameters can be assumed as the related vector:

$$\mathbf{x} = \mathbf{x}\,(t)\,, \qquad (5.2.4)$$

where $\mathbf{x}$ and $\mathbf{x}\,(t)$ are relatively the vector of controlled parameters envisaged for the considered automation object and the related vector function representing the dependencies of these parameters on the time.

Of course, the senses and the count of the controlled parameters (5.2.4) are defined by the nature of the processes inherent in the considered automation object, as well as by purposes of the related automation system in each particular case. The general sense of the vector (5.2.4) is due to their usage in the formal definition required to build the automation object's mathematical model, which is imagined as the relation between the controlled parameters and the controlling parameters represented relatively by the vectors (5.2.4) and (5.2.2) for this automation object. It is natural that the most general representation of the relation between the vectors (5.2.4) and (5.2.2) is through the ordinary differential equations, but it is impossible to give any uniform recommendations on how to build these equations because their view is sensitive to the particularities of the considered automation object. At the same time, from the theoretical point of view, it is accessible that the velocities of the controlled parameters (5.2.4) are the results of the state and their changes' velocities, the controlling parameters and their changes' velocities:

$$\frac{d\mathbf{x}}{dt} = \dot{\mathbf{x}}\left(t, \mathbf{q}, \frac{d\mathbf{q}}{dt}, \frac{d^2\mathbf{q}}{dt^2}, \ldots; \mathbf{u}(t), \frac{d\mathbf{u}(t)}{dt}, \frac{d^2\mathbf{u}(t)}{dt^2}, \ldots\right), \quad \mathbf{x}(t_0) = \mathbf{x}_0, \tag{5.2.5}$$

where $\dot{\mathbf{x}}(t, \mathbf{q}, \ldots; \mathbf{u}(t), \ldots)$ is the given vector function defining the velocities of changes of the controlled parameters depending on the time, the current state, and the controlling influences in the considered automation object; t_0 and x_0 are the given initial time and the given related vector of the initial values of the controlled parameters envisaged in the considered automation object assumed to consider their researched operation mode.

The principal from the theoretical points of view in the differential equations (5.2.5) is that the velocities of changes of the controlled parameters are not dependent on these parameters, because the changes of the controlled parameters are possible only due to changes of the state and controls of the automation object, but not due to changes of these controlled parameters themself. So, the generalized view of the mathematical model of the automation object is assumed in the form of the coupled systems (5.2.3) and (5.2.5) of the ordinary differential equations with the related initial conditions, and such mathematical model can be schematically imagined as it is shown on Figure 5.1. As it is shown by the schematically imagination on Figure 5.1, the principal difficulties of the mathematical modeling of automation objects are due to requirements of the consideration of the intermediate parameters (5.2.1) defining the state of the processes inherent for the considered automation object. Of course, it is theoretically possible to exclude the parameters (5.2.1) of the processes inherent for the considered

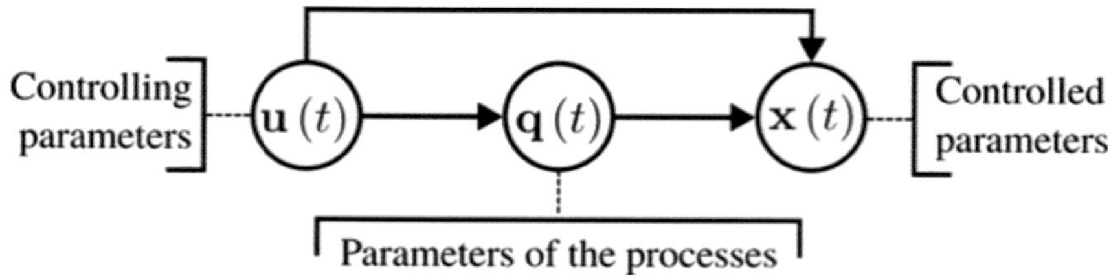

Figure 5.1 Processes in the automation object.

automation object from the system of the differential equations (5.2.3) and (5.2.5) to have the mathematical model of this considered automation object defined in the terms of the controlling (5.2.2) and controlled (5.2.4) parameters only, but it is practically possible only for some particular cases. The principal difficulties for excluding the processes' state parameters (5.2.1) from the mathematical models (5.2.3) and (5.2.5) of automation objects are due to the different dimensions of the vectors (5.2.1), (5.2.3) and (5.2.4) so that, the dimension of the vector (5.2.1) is often significantly bigger than the dimensions of the vectors (5.2.1) and (5.2.3). Nevertheless, the system (5.2.3), and (5.2.5) of ordinary differential equations with the related initial conditions gives the most general representation for the mathematical model of automation objects, and due to such representation, it is clearly seen how the inherent processes form the properties of the researched object exactly as the automation object. While the initial value problems (5.2.3) are considered and are widely researched in mathematical physics and related mathematical fields, the initial value problems (5.2.5) are not widely researched at present, because automation engineering is traditionally divided into the different fields related to the particular applications like the automotive automation, the thermal automation, and others having the own particular mathematical models of the automation objects, so that the generalized universal approaches allowing to cover the construction of the mathematical models for different automation objects are out of the attention, although such general approaches are important to provide automation systems with additional energy and resource-saving functions, as it is shown further.

5.2.3 Linearized mathematical models of automation objects

One of the principal purposes of the automation systems is to support the required values of the controlled parameters of automation objects during the operational time, and the linearized mathematical models of automation objects are used to design the suitable automatic governors to do it. So, it is important to provide the linearization of the generally represented mathematical models (5.2.3) and (5.2.5) of automation objects.

To design the suitable automatic governors, the linearized mathematical models of automation objects with the controlling and controlled parameters (5.2.3) and (5.2.5) must be presented using the following linear ordinary differential equations:

$$\mathbf{A}_n \frac{d^n \mathbf{x}}{dt^n} + \ldots + \mathbf{A}_1 \frac{d\mathbf{x}}{dt} + \mathbf{A}_0 \mathbf{x} = \mathbf{B}_m \frac{d^m \mathbf{u}}{dt^m} + \ldots + \mathbf{B}_1 \frac{d\mathbf{u}}{dt} + \mathbf{B}_0 \mathbf{u}, \quad (5.2.6)$$

where n is the order of the differential equation of the linearized mathematical model of automation object; m is the integer value defining the order of the accelerations of the controlling parameters involved to the differential equation of the linearized mathematical model of automation object; $\mathbf{A}_0$, $\mathbf{A}_1$,..., $\mathbf{A}_n$ and $\mathbf{B}_0$, $\mathbf{B}_1$,..., $\mathbf{B}_m$ are the matrices defining the ordinary differential equations, whose represent the linearized mathematical model of the automation object.

The principal task is to build the linear differential equations (5.2.6) as the result of the linearization procedures for the automation object represented through the mathematical model of the general view (5.2.3) and (5.2.5). Of course, it is theoretically possible to make some analytical equivalent transformations of the differential equations from the generalized mathematical model (5.2.3) and (5.2.5) of the automation object to reduce these generalized equations to the linear ordinary differential equations (5.2.6). At the same time, such analytical transformations are practically possible only for specific cases of automation objects, and they are not practically possible in general. So, the general way to build the linearized ordinary differential equations (5.2.6) from the mathematical model (5.2.3) and (5.2.5) of the automation objects is related not to the analytical transformations, but only to the usage of the numerical procedures. The parametric identification is proposed as the theoretical foundation for these noted numerical procedures to build the linearized ordinary differential equations (5.2.6) from the mathematical model (5.2.3) and (5.2.5) of the automation objects, and such an approach is schematically shown in Figure 5.1. Such parametric identification procedure is illustrated in Figure 5.2 and directed to build the linearized ordinary differential equations (5.2.6) from the mathematical model (5.2.3) and (5.2.5) of the automation objects can be represented by the following mapping:

$$\mathbf{u}(t), \ \mathbf{x}(t) \xrightarrow{I_P} \mathbf{A}_n, \ldots, \mathbf{A}_1, \mathbf{A}_0, \mathbf{B}_m, \ldots, \mathbf{B}_1, \mathbf{B}_0, \quad (5.2.7)$$

where I_P is the method for the parametric identification procedure.

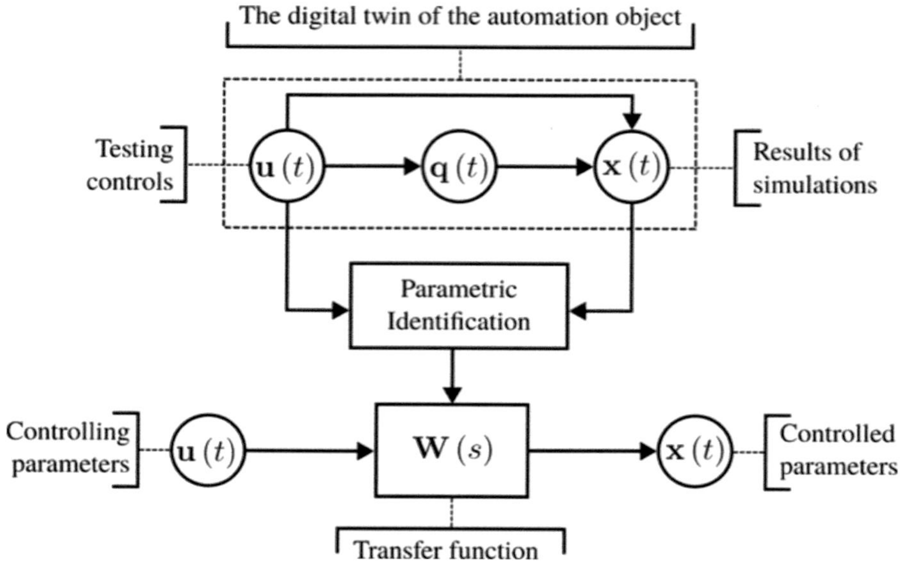

Figure 5.2 The mathematical modeling of the automation object.

The principal idea of the implementation of the parametric identification (5.2.7) is in the usage of the controlling parameters $\mathbf{u}(t)$ and the related controlled parameters $\mathbf{x}(t)$ from the results of the computer simulations through the implementation of suitable numerical methods to solve the coupled system (5.2.3) and (5.2.5) of the ordinary differential equations with the corresponded initial conditions representing the considered automation object. The mathematical model (5.2.3) and (5.2.5) is imagined in this approach as the digital twin (Figure 5.2) of the considered automation object, but the computer simulations are imagined as the testing of the digital twin of this automation object. Of course, the testing controls $\mathbf{u}(t)$ allowing the results of simulations $\mathbf{x}(t)$ must be suitably chosen to make the parametric identification (5.2.7) as shown in Figure 5.2, but it is a separate problem not for this research. In any case, the consideration of the identification problem (5.2.7) allows to have the transfer function associated with the linear model (5.2.6) in the following view:

$$\mathbf{W}(s) = \left(\mathbf{A}_n s^n + \ldots + \mathbf{A}_1 s + \mathbf{A}_0\right)^{-1} \left(\mathbf{B}_m s^m + \ldots + \mathbf{B}_1 s + \mathbf{B}_0\right), \quad (5.2.8)$$

where $\mathbf{W}(s)$ is the transfer function related to the linearized mathematical model of the considered automation object (5.2.8).

It is necessary to highlight that the transfer function (5.2.8) represents the properties of the processes, inherent to the considered automation, object through the results of testing for the related digital twin (Figure 5.2).

5.3 Energy and Resource Saving for Transient Modes

The transient modes are defined as the modes related to a change in the controlled parameters of an automation object from one current state to a new steady-state value during the operating time, and the development of energy and resource-saving automated controls for such transient modes is the relatively separate problem of automation engineering. In this research, the development of energy and resource-saving automated controls for transient modes on automation objects is reduced to consideration of the related problems of the optimal control theory.

5.3.1 Formulations of related problems about optimal control

The considered automation object is imagined through their mathematical model represented in the general view (5.2.3) and (5.2.5). While the transient operating mode is considered, the initial time t_0 and the related initial vectors q_0 and x_0 in the mathematical model (5.2.3) and (5.2.5) are associated with some current state of the automation object. The researched transient modes in the considered automation object are defined by the initial conditions in the mathematical model (5.2.3) and (5.2.5) and by the final conditions of the following view:

$$\mathbf{x}(t_f) = \mathbf{x}_f, \tag{5.3.1}$$

where t_f and $\mathbf{x}_f$ are the given final time and the given related vector of the final values of the controlled parameters envisaged in the considered automation object assumed to consider their transient mode.

Thus, the transient mode of the considered automation object is defined by means of the mathematical model (5.2.3) and (5.2.5), and the final conditions (5.3.1), so that the energy and resource-saving automated controls must be developed for these defined transient modes. Of course, the current and the final conditions defining the researched transient modes must be achieved principally for the considered automation object including without any damages or other inherent restrictions. It is assumed that the initial conditions from the mathematical model (5.2.3) and (5.2.5), and the final condition (5.3.1) defining the researched transient modes can be achieved for the considered automation object.

It is clear that the controls (5.2.2) during any of the operational modes must be limited to prevent any damage to the automation object, and such limitation of the controls can be imagined as the attachment of the controls to the related functional space:

$$\mathbf{u} \in U \quad \Leftrightarrow \quad \|\mathbf{u}\|_U \leq [u]\,, \tag{5.3.2}$$

where U is the normed space of the permissible controls of the considered automation object; $\|\bullet\|_U$ is the norm introduced in the space U; $[u] > 0$ is the given value defining the permissible control of the considered automation object.

The consideration of the transient modes defined by the mathematical model (5.2.3) and (5.2.5), and the final conditions (5.3.1) leads to the possible controls, whose form the subspace in the space $U$of the permissible controls:

$$\mathbf{u} \in U_f \quad \Leftrightarrow \quad \mathbf{x}(t_0) = \mathbf{x}_0,\ \mathbf{x}(t_f) = \mathbf{x}_f, \tag{5.3.3}$$

where $U_f \subset U$ is the subspace of possible controls in the space of permissible controls.

The energy is assumed further as the particular type of the resources to reduce the formulations into the short form in terms of resource saving. It is assumed in this research that the different types of resources used during the operation of the considered automation object can be defined by the related vector:

$$\mathbf{r} = \mathbf{r}(t)\,, \tag{5.3.4}$$

where $\mathbf{r}$ and $\mathbf{r}(t)$ are the vector of amounts of different types of resources already consumed during the operation of the considered automation object and the related vector function depending on the time and representing of the already used resources at the correspondent time moments.

The vector introduced in the view (5.3.4) is shortly named further as the vector of used resources in the considered automation object. The velocity of the usage of the resource during the operational time is defined by the current state of the processes inherent for the considered automation object, so the already used resources are defined by the following ordinary differential equations and the related initial conditions

$$\frac{d\mathbf{r}}{dt} = \dot{\mathbf{r}}\left(t, \mathbf{q}, \frac{d\mathbf{q}}{dt}, \frac{d^2\mathbf{q}}{dt^2}, \ldots; \mathbf{u}(t)\,, \frac{d\mathbf{u}(t)}{dt}, \frac{d^2u(t)}{dt^2}, \ldots,\right), \quad \mathbf{r}(t_0) = \mathbf{r}_0, \tag{5.3.5}$$

where $\dot{\mathbf{r}}(t, \mathbf{q}, \ldots; \mathbf{u}(t)\,, \ldots)$ is the given vector function defining the velocities of the resources' usage depending on the time, the current state, and the

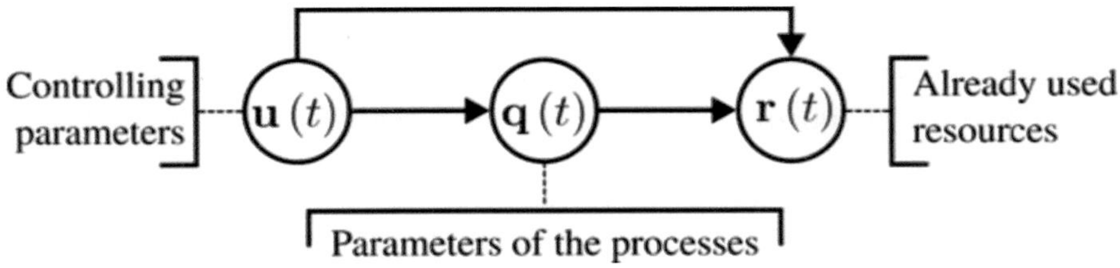

Figure 5.3 The mathematical modeling of used resources in the automation object.

controlling influences in the considered automation object; t_0 and r_C are the given initial time and the given related vector of the already used resources before the initial time in the considered automation object.

It is clear that the view of the resource usage velocities defining the differential equations (5.3.4) is sensitive to the type of the considered resources and to the particularities of the considered automation object. In any case, the relations (5.3.5) represent the mathematical model of the resources' usage in the considered automation object so that this mathematical model (5.3.5) with the mathematical model (5.2.3) of the processes allows an estimate of the used resources in the considered automation object during the operational time, as it is shown in Figure 5.3. The mathematical model (5.2.3), (5.2.5), and (5.3.5) of the resource usage in the automation object can be imagined as the mapping of the following view:

$$\mathbf{u}(t) \xrightarrow{M_N} \mathbf{r}(t) \quad \Rightarrow \quad \mathbf{r} = \mathbf{r}(t; \mathbf{u}(t)), \tag{5.3.6}$$

where M_N is the numerical method suitable to integrate the system (5.2.3), (5.2.5), and (5.3.5) of the ordinary differential equations with the related initial conditions, representing the mathematical model of the resource usage in the considered automation object; $\mathbf{r}(t; \mathbf{u}(t))$ is the vector function representing the solution of the system (5.2.3), (5.2.5), and (5.3.5) of the ordinary differential equations with the related initial conditions for the given controlling parameters represented by means the corresponded vector $\mathbf{u}(t)$.

The mapping (5.3.6) allows to introduce formally the resource-saving control for the researched transient mode of the considered automation object

$$\mathbf{u}_R(t) \in U_f : \quad \forall \mathbf{u}(t) \in U_f \Rightarrow \| \mathbf{r}(t_f; \mathbf{u}(t)) \| \geq \| \mathbf{r}(t_f; \mathbf{u}_R(t)) \|, \tag{5.3.7}$$

where $\mathbf{u}_R(t)$ is the resource-saving control providing the required transient mode for the considered automation object; $\|\bullet\|$ is some norm of the vector.

The resource-saving control introduced in the view (5.3.7) is in agreement with the optimal control problem as it is considered in the well-known article [39]. At the same time, in the article [39] the particular case is considered because it is envisaged that the differential equations representing the relation between controlling and controlled parameters are given in the direct view without the intermediate unknowns defining the processes inherent for the considered automation object. Although article [39] gives the principal results about optimal controls, it is required to research the continuation of these results to the cases of the indirectly defined differential equations representing the relation between controlling and controlled parameters, as it is in this research through the intermediate unknowns representing the parameters of the processes inherent for the considered automation object (Figure 5.1). Such fundamental properties of the resource-saving controls represented through the related optimal controls are important, but the purposes of this research are related to the application-aided approaches, so only the approaches based on the numerical methods to construct the resource-saving controls (3.7) are considered on this research suitable to solve the automation engineering applied problems.

5.3.2 The related optimization problem for cases of transient modes

The principal idea of the approaches to build the resource-saving controls (5.3.7) is in the reduction of the related optimal control problem to the equivalent finite-dimensional optimization problem and in the use of suitable numerical methods to solve it, as discussed in the chapter [40]. Such approaches are in agreement with the general purposes of this research in providing the computer-aided automation engineering analysis for the designing of automation systems with energy efficiency and resource-saving properties.

The controlling parameters (5.2.2) are the unknowns in the optimal control problem (5.3.7). Let us consider the controlling parameters of the considered automation object in the following view:

$$\mathbf{u}(t) = \mathbf{a}_0 + \mathbf{a}_1 t + \mathbf{a}_2 t^2 + \ldots + \mathbf{a}_n t^n, \tag{5.3.8}$$

where n is the count of the accounted time modes of the controlling parameters; $\mathbf{a}_0, \mathbf{a}_1, \mathbf{a}_2, \ldots, \mathbf{a}_n$ are the vectors of the parameters defining the related time modes involved to the controlling parameters.

The same view of the relation (5.3.8) with Taylor's series allows us to assume that for the big value n, it is possible to cover wide classes of the considered controls. At the same time, the small value n in the relation (5.3.8) suitable for applied tasks significantly restricts the classes of the considered controls. Nevertheless, the controlling parameters (5.3.8) are more useful for automation engineering applications, exactly due to the significant restrictions on the classes of the accounted controls. Indeed, the controls in the automation object can be formed practically as the combination of the different time modes, and the count of these time modes is restricted by the small numbers. So, the representation (5.3.8) of the controlling parameters significantly narrows the optimal control problem from the theoretical point of view, but such representation (5.3.8) is in full agreement with the applied problems considered in the fields of automation engineering. Taking into account the relation (5.3.8), the mathematical model (5.2.3), (5.2.5), and (5.3.5) are reduced to the following view:

$$\frac{d\mathbf{q}}{dt} = \dot{\mathbf{q}}\left(t, \mathbf{q}; \mathbf{a}_0, \mathbf{a}_1, \mathbf{a}_2, \ldots, \mathbf{a}_n\right), \quad \mathbf{q}\left(t_0\right) = \mathbf{q}_0, \tag{5.3.9}$$

$$\frac{d\mathbf{x}}{dt} = \dot{\mathbf{x}}\left(t, \mathbf{q}, \frac{d\mathbf{q}}{dt}, \frac{d^2\mathbf{q}}{dt^2}, \ldots; \mathbf{a}_0, \mathbf{a}_1, \mathbf{a}_2, \ldots, \mathbf{a}_n\right), \quad \mathbf{x}\left(t_0\right) = \mathbf{x}_0, \tag{5.3.10}$$

$$\frac{d\mathbf{r}}{dt} = \dot{\mathbf{r}}\left(t, \mathbf{q}, \frac{d\mathbf{q}}{dt}, \frac{d^2\mathbf{q}}{dt^2}, \ldots; \mathbf{a}_0, \mathbf{a}_1, \mathbf{a}_2, \ldots, \mathbf{a}_n\right), \quad \mathbf{r}\left(t_0\right) = \mathbf{r}_0. \tag{5.3.11}$$

The mathematical model (5.3.9)–(5.3.11) allows to estimate controlled parameters and the already used resource for the considered automation object for the given parameters defining the time modes involved to the controlling parameters, and such estimations can be symbolically represented by the following relations:

$$\mathbf{x} = \mathbf{x}\left(t; \mathbf{a}_0, \mathbf{a}_1, \mathbf{a}_2, \ldots, \mathbf{a}_n\right), \tag{5.3.12}$$

$$\mathbf{r} = \mathbf{r}\left(t; \mathbf{a}_0, \mathbf{a}_1, \mathbf{a}_2, \ldots, \mathbf{a}_n\right). \tag{5.3.13}$$

Of course, relations (5.3.12) and (5.3.13) cannot be obtained in the direct view in the general case, but the usage of the suitable numerical methods to solve the system (5.3.9)–(5.3.11) of the ordinary differential equations with the related initial conditions allows to define the vectors **x** and **r** for the given values of the parameters $\mathbf{a}_0, \mathbf{a}_1, \mathbf{a}_2, \ldots, a_n$, so that the relations (5.3.12) and (5.3.13) are just the notations for such possibilities. The optimization problem involves usually, the objective function, the equality constraints, and the inequality constraints, as highlighted in the chapter [40]. To approximately build the resource-saving control (5.3.7) defined by the related optimal control problem for the researched transient mode of the considered automation object, it is necessary to solve the optimization problem with the objective function defined by the relation (5.3.13), equality constraints defining the possible controls through the final condition (5.3.3), and the inequality constraints defined by the definition (5.3.2) of the permissible controls. So, taking into account the relations (5.3.8), (5.3.12), and (5.3.13), the equivalent optimization problem to find the parameters $\mathbf{a}_0, \mathbf{a}_1, \mathbf{a}_2, \ldots, \mathbf{a}_n$ corresponding to the resource-saving automated controls for the researched transient mode of the considered automation object can be formulated in the following view:

$$F\left(\mathbf{a}_0, \mathbf{a}_1, \mathbf{a}_2, \ldots, \mathbf{a}_n\right) \rightarrow \min, \tag{5.3.14}$$

$$\mathbf{x}\left(t_f; \mathbf{a}_0, \mathbf{a}_1, \mathbf{a}_2, \ldots, \mathbf{a}_n\right) = \mathbf{x}_f, \tag{5.3.15}$$

$$\left\|\mathbf{a}_0 + \mathbf{a}_1 t + \mathbf{a}_2 t^2 + \ldots + \mathbf{a}_n t^n\right\|_U \leq [u], \tag{5.3.16}$$

where $F\left(\mathbf{a}_0, \mathbf{a}_1, \mathbf{a}_2, \ldots, \mathbf{a}_n\right) = \left\|\mathbf{r}\left(t_f; \mathbf{a}_0, \mathbf{a}_1, \mathbf{a}_2, \ldots, \mathbf{a}_n\right) - \mathbf{r}_0\right\|$ is the objective function of the optimization problem.

The view (5.3.14)–(5.3.16) of the equivalent optimization problem is the same as the one considered in the chapter [40], so all well-known engineering optimization methods can be used. At the same time, it is necessary to highlight that the objective function (5.3.13) used to define the optimality condition (5.3.14) is not given in direct view in a general case, but this objective function can be computed for their given arguments through the numerical solving of the system (5.3.9)–(5.3.11) of the ordinary differential equations with the related initial conditions. Nevertheless, a lot of well-known optimization methods [40] require only the evaluation possibility of the objective function, so the indirectly defined objective function (5.3.13) does not restrict the opportunities to solve such formulated optimization problem, but solving such a problem requires a lot of computing time. Thus, the construction of the energy and resource-saving controls for the transient modes of the considered automation object is reduced to the equivalent constrained optimization problem with equality and inequality constraints, which can be solved using any known engineering optimization methods [40].

5.4 Energy and Resource Saving PID-Governors

The steady-state modes are assumed as the modes related to the given values of controlled parameters of an automation object, and the development of energy and resource-saving automated controls for such steady-state modes is a relatively separate problem in automation engineering, which is related to the construction of the automatic governors. The PID-governors are the most used in automation engineering, so their improvement is discussed in the context of offering the parameters choosing approaches allowing to provide energy and resource-saving on the steady-state modes of automation objects so that these approaches are based on reducing equivalent constrained finite-dimensional optimization problem suitable to use the related known numerical methods.

5.4.1 The mathematical modeling of PID-governors

The principal idea of the usage of automatic PID-governors is in the correcting of the controlling parameters to provide almost stable values of the controlled parameters of the considered automation object during the long-term operational time on the related steady-state mode. While the steady-state mode is researched, the introduced values (5.2.4) are imagined not as the own controlled parameters, but as the differences between the current and required values of the controlled parameters inherent for the considered automation object. Due to this assumption, the correcting of the control parameters by means of the PID-governor takes the form:

$$\mathbf{u}(t) = -\mathbf{K}_P\mathbf{x}(t) - \mathbf{K}_I\int_0^t \mathbf{x}(\tau)\,d\tau - \mathbf{K}_D\frac{d}{dt}\mathbf{x}(t), \tag{5.4.1}$$

where $\mathbf{K}_P$, $\mathbf{K}_I$, and $\mathbf{K}_D$ are the matrices involving the parameters of the used PID-governor, defining proportional, integral, and differential parts of it.

To analyze the influence of the PID-governor (5.4.1) on the automation object indirectly represented by the system (5.2.3) and (5.2.5) of the ordinary differential equations, it is necessary to introduce a new intermediate variable:

$$\frac{d\mathbf{y}}{dt} = \mathbf{x}, \quad \mathbf{y}(t_0) = \mathbf{0}, \tag{5.4.2}$$

where $\mathbf{y} = \mathbf{y}(t)$ is the new intermediate variable required to account the integral part of the PID-governor; $\mathbf{0}$ is the zeroes vector having the related dimension.

The introduced new variable (5.4.2) allows representing the correcting (5.4.1) of the controlling parameters using the PID-governor in the more suitable form:

$$\mathbf{u}(t) = -\mathbf{K}_P\mathbf{x}(t) - \mathbf{K}_I\mathbf{y}(t) - \mathbf{K}_D\frac{d}{dt}\mathbf{x}(t), \tag{5.4.3}$$

So, the mathematical model of the automation object with the envisaged PID-governor has the view of the system of differential equations (5.2.3),

(5.2.5), (5.4.2), and the complementary relation (5.4.3). It demonstrates that the involvement of the PID-governor significantly changes the properties of the automation object. Indeed, the integral part of the PID-governor leads to an increase in the number of differential equations representing the mathematical model of the automation object due to the additional equation (5.4.2). The differential part of the PID-governor leads to changes in the differential equations of the mathematical model (5.2.3) and (5.2.5) representing the processes and the controlled parameters inherent for the automation object, and such changes are especially noticeable in the cases of nonlinear differential equations representing the automation object. It is necessary to highlight that the simulation of the processes in the automation objects with the envisaged PID-governor is important because the corrections made by the PID-governors to the controlling parameters significantly influence the energy and resource consumption of the automation object during the operational time.

5.4.2 PID-governors in the linearized mathematical models

The parameters $\mathbf{K}_P$, $\mathbf{K}_I$, and $\mathbf{K}_D$ of the PID-governor must be chosen to provide automatically the stability of the required values of the controlled parameters during the operational time of the considered automation object on the corresponding steady-state mode. Such stability can be provided by means the usage of the linearized differential equations (5.2.6) of the considered automation object. To do it, the new variable introduced in the view (5.4.2) is used to transform the linear differential equations (5.2.6) and the controls (5.4.1) to the more suitable form:

$$\mathbf{A}_n \frac{d^{n+1}\mathbf{y}}{dt^{n+1}} + \ldots + \mathbf{A}_1 \frac{d^2\mathbf{y}}{dt^2} + \mathbf{A}_0 \frac{d\mathbf{y}}{dt} = \mathbf{B}_m \frac{d^m\mathbf{u}}{dt^m} + \ldots + \mathbf{B}_1 \frac{d\mathbf{u}}{dt} + \mathbf{B}_0\mathbf{u}, \quad (5.4.4)$$

$$\mathbf{u}(t) = -\mathbf{K}_D \frac{d^2}{dt^2}\mathbf{y}(t) - \mathbf{K}_P \frac{d}{dt}\mathbf{y}(t) - \mathbf{K}_I \mathrm{y}(t). \quad (5.4.5)$$

The controls (5.4.5) related to the PID-governor must be substituted to the differential equations (5.4.4), and the following result for the case $n > m$ has the following view:

$$\begin{aligned} &\mathbf{A}_n \tfrac{d^{n+1}\mathbf{y}}{dt^{n+1}} + \ldots \\ &+ (\mathbf{A}_{m+2} + \mathbf{B}_m\mathbf{K}_D) \tfrac{d^{m+2}\mathbf{y}}{dt^{m+2}} + (\mathbf{A}_{m+1} + \mathbf{B}_m\mathbf{K}_P) \tfrac{d^{m+1}\mathbf{y}}{dt^{m+1}} + (\mathbf{A}_m + \mathbf{B}_m\mathbf{K}_I) \\ &\tfrac{d^m\mathbf{y}}{dt^m} + + \ldots + (\mathbf{A}_1 + \mathbf{B}_0\mathbf{K}_D) \tfrac{d^2\mathbf{y}}{dt^2} + (\mathbf{A}_0 + \mathbf{B}_0\mathbf{K}_P) \tfrac{d\mathbf{y}}{dt} + \mathbf{B}_0\mathbf{K}_I\mathbf{y} = 0. \end{aligned} \quad (5.4.6)$$

The solution of the system (4.6) of the linear ordinary differential equations can be represented using the matrix exponential:

$$\mathbf{y}(t) = e^{\Lambda t}\mathbf{C}_0, \tag{5.4.7}$$

where $\mathbf{\Lambda}$ is the matrix of eigenvalues;C_0 is the vector related to the initial conditions required for the system (5.4.6).

The substitution of the solution represented in the view (5.4.7) to the system (5.4.6) of linear differential equations leads to the characteristic equation in the following matrix form:

$$\det \left(\mathbf{A}_n\mathbf{\Lambda}^{n+1} + \ldots \right.$$
$$+ \left(\mathbf{A}_{m+2} + \mathbf{B}_m\mathbf{K}_D\right)\mathbf{\Lambda}^{m+2} + \left(\mathbf{A}_{m+1} + \mathbf{B}_m\mathbf{K}_P\right)\mathbf{\Lambda}^{m+1} + \left(\mathbf{A}_m + \mathbf{B}_m\mathbf{K}_I\right)\mathbf{\Lambda}^{m} +$$
$$\left. + \ldots + \left(\mathbf{A}_1 + \mathbf{B}_0\mathbf{K}_D\right)\mathbf{\Lambda}^2 + \left(\mathbf{A}_0 + \mathbf{B}_0\mathbf{K}_P\right)\mathbf{\Lambda} + \mathbf{B}_0\mathbf{K}_I\right) = 0. \tag{5.4.8}$$

The characteristic equation (5.4.8) allows us to find the matrix Λ of eigenvalues. Of course, the matrix Λ of eigenvalues defined from the characteristic equation (5.4.8) has dependences on the automation object properties defined by means of the matrices $\mathbf{A}_0$, $\mathbf{A}_1$,..., $\mathbf{A}_n$ and $\mathbf{B}_0$, $\mathbf{B}_1$,..., $\mathbf{B}_m$, but it has dependencies also on the parameters $\mathbf{K}_P$, $\mathbf{K}_I$ and $\mathbf{K}_D$ of the PIG-governor. It is assumed that the automation object properties are given, but the parameters of the PID-governor must be defined for such given automation object. So, for the given matrices $\mathbf{A}_0$, $\mathbf{A}_1$,..., $\mathbf{A}_n$ and $\mathbf{B}_0$, $\mathbf{B}_1$,..., $\mathbf{B}_m$ defining the automation object, the matrix $\mathbf{\Lambda}$ of eigenvalues defined from the characteristic equation (5.4.8) can be imagined as having dependencies on the parameters of the PID-governor, and it can be represented symbolically in the view:

$$\mathbf{\Lambda} = \mathbf{\Lambda}\left(\mathbf{K}_P, \mathbf{K}_D, \mathbf{K}_I\right), \tag{5.4.9}$$

where $\mathbf{\Lambda}\left(\mathbf{K}_P, \mathbf{K}_D, \mathbf{K}_I\right)$ is the relation defined by the characteristic equation (5.4.6) for the given matrices $\mathbf{A}_0$, $\mathbf{A}_1$,..., $\mathbf{A}_n$ and $\mathbf{B}_0$, $\mathbf{B}_1$,..., $\mathbf{B}_m$.

While the (5.2.4) is assumed as the differences between the current and required values of the controlled parameters, the stability of the researched steady-state values (5.4.2) can be imagined as the zero limits:

$$\lim_{t\to\infty} \mathbf{y}(t) = 0. \tag{5.4.10}$$

To provide the limits (5.4.10), the following conditions must be satisfied:

$$Re\Lambda_{ij}\left(\mathbf{K}_P, \mathbf{K}_D, \mathbf{K}_I\right) < 0, \quad i, j = 1, 2, \ldots, n, \tag{5.4.11}$$

where $\Lambda_{ij}\left(\mathbf{K}_P, \mathbf{K}_D, \mathbf{K}_I\right)$ are the elements of the matrix (5.4.9) of eigenvalues.

It is necessary to highlight again that the conditions (5.4.11) are for the given automation object defining through the linear model (5.2.6) by the matrices $\mathbf{A}_0$, $\mathbf{A}_1,\ldots$, $\mathbf{A}_n$ and $\mathbf{B}_0$, $\mathbf{B}_1,\ldots$, $\mathbf{B}_m$ depending on the design of this automation object. So, the conditions (5.4.11) for the parameters of the PID-governor are dependent of the design of the automation object through the given matrices $\mathbf{A}_0$, $\mathbf{A}_1,\ldots$, $\mathbf{A}_n$ and $\mathbf{B}_0$, $\mathbf{B}_1,\ldots$, $\mathbf{B}_m$ which are used to build these conditions. This circumstance allows to see that the design of the automation object influences the parameters of the PID-governor, and it is possible and reasonable to account for this during the development of the automation object to improve opportunities for automatic system implementations.

5.4.3 PID-governors with energy- and resource-saving properties

The conditions (5.4.11) allow the set of the possible values for the parameters $\mathbf{K}_P$, $\mathbf{K}_I$, and $\mathbf{K}_D$ of the PIP-governor, so the principal idea is to choose the values of these parameters providing the energy and resource-saving automatic control for the researched steady-state mode of the considered automation object. To choose such values of the parameters $\mathbf{K}_P$, $\mathbf{K}_I$, and $\mathbf{K}_D$ of the PID-governor, the related optimization problem is formulated further.

To define the parameters $\mathbf{K}_P$, $\mathbf{K}_I$, and $\mathbf{K}_D$ of the PID-governor providing the energy and resource-saving automatic control for the researched steady-state mode of the considered automation object, it is necessary to consider the energy and resource consumptions during the operational time on this steady-state mode. Such consideration is based on the numerical solving of the mathematical model (5.2.3), (5.2.5), (5.3.4), and (5.4.2) taking into account the automatic control (5.4.3) related to the PIG-governor. For all given parameters defining the automation object, but for variations of the values of the parameters $\mathbf{K}_P$, $\mathbf{K}_I$, and $\mathbf{K}_D$ of the PID-governor the results of this numerical solving can be symbolically represented by the following relations:

$$\mathbf{r} = \mathbf{r}\left(t; \mathbf{K}_P, \mathbf{K}_D, \mathbf{K}_I\right), \quad t \geq t_0. \tag{5.4.12}$$

Of course, the relations (5.4.12) cannot be obtained in the direct view in general case, but the usage of the suitable numerical methods to solve the system (5.2.3), (5.2.5), (5.3.4), and (5.4.2) of the ordinary differential equations with the related initial conditions and taking into account with the relation (5.4.3) allows to define the vector r of the consumed energy and resources for the given values of the parameters $\mathbf{K}_P$, $\mathbf{K}_I$, and $\mathbf{K}_D$ of the envisaged PID-governor, so that the relations (5.4.12) are just the notations for such possibilities. Nevertheless, the relation (5.4.12) can be used to define the objective function:

$$F\left(\mathbf{K}_P, \mathbf{K}_D, \mathbf{K}_I\right) \rightarrow \min, \tag{5.4.13}$$

where $F\left(\mathbf{K}_P, \mathbf{K}_D, \mathbf{K}_I\right) = \left\|\mathbf{r}\left(t_0 + t_s; \mathbf{K}_P, \mathbf{K}_D, \mathbf{K}_I\right) - \mathbf{r}_0\right\|$ is the objective function of the optimization problem; t_s is the duration of the researched steady-state mode of the considered automation object.

To design the PID-governors providing energy and resource efficiency during the operation of the considered automation object, it is necessary to solve the optimization problem formulated for the objective function (5.4.13) and the inequality constraints (5.4.11). The formulation of this optimization problem (5.4.13) and (5.4.11) is in full agreement with the known results in the book [40], so it is possible to use the known numerical methods for engineering optimization. At the same, time, it is reasonable to highlight that the objective function (5.4.13) and the inequality constraints (5.4.11) are given in the indirect form through the complicated auxiliary problems, so that it is possible to use only the numerical methods, and it is required a lot of computing time to solve the related optimization problem. Thus, the construction of the energy and resource-saving PID-governors for the steady-state modes of the considered automation object is reduced to the equivalent constrained optimization problem with the inequality constraints, which can be solved by means of any known engineering optimization methods [40].

5.5 Energy-Saving Automation for Electric Drives

Electric drives are widely used for industrial, household, transportation, and other purposes due to their power characteristics, and zero carbon emission; besides, they are suitable for Internet of Things applications due to simple control through electrical and electronic devices, which can be easily conjugated with computers. The design and technologies of electric machines are close to the possibilities limited by the thermodynamical efficiency so it becomes difficult each year to improve such machines more. At the same time, the improvement of automation systems for electric drives allows energy- and resource-saving during the operational time even for the existing electrical machines, and such possibilities are shown and discussed further, including the examples of the proposed general theory.

5.5.1 Direct current electric drives as the automation objects

The electric drives are used to provide the rotational moment from voltage sources to move working parts, as schematically shown in Figure 5.4. An electric drive must necessarily involve an electric motor, and it usually involves the reduction gear (Figure 5.4) to agree on the angular velocity of a rotor of the electric motor with the angular velocity required to move working parts, and besides, to increase the rotational moment of the electric motor, so that the following relation between the angular velocities:

$$\omega_e = n_G \omega, \tag{5.5.1}$$

where ω_e is the angular velocity of the rotor of the electric motor; ω is the angular velocity of the output shaft of the reduction gear, which is provided by the electric drive to move the working parts; $n_G > 1$ is the reduction ratio.

The electric motor is the principal constituent defining significantly the properties of the electric drive. A lot of electric motor types are possible to use, but the electric drive with the direct current motor is considered in this research due to their wide implementations in different applications, especially for the Internet of Things. The electric drive is actually the electromechanical system, so to define their state, it is necessary to define the state of all their existing electrical and mechanical constituents. Due to this circumstance and the inherent relation (5.5.1), it is required to use at least two parameters to define the state of the electric drive, so that one of them

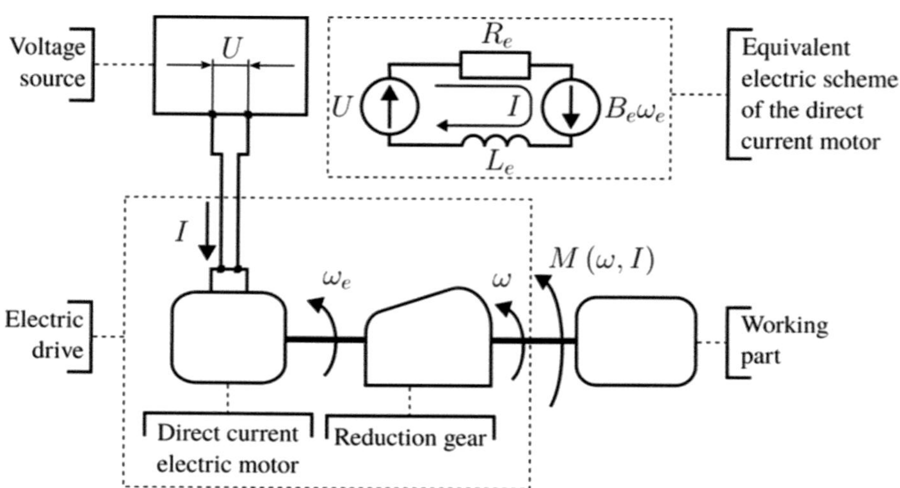

Figure 5.4 Schematic representations about an electric drive.

is to define the electrical constituents, but the second of them is to define the mechanical constituents so that the generally introduced vector (5.2.1) related with such schematization of the electric drive has the following view:

$$\mathbf{q} = \begin{pmatrix} q_1 \\ q_2 \end{pmatrix}, \quad q_1 = \omega, \; q_2 = I, \tag{5.5.2}$$

where q_1 and q_2 are the state parameters of the electric drive defining the state of the mechanical and electrical constituents of the electric drive; I is the electric current in the winding of the rotor of the electric motor.

The rotational moment provided by the electric drive is due to the electric voltage, so it is natural and reasonable to consider this supplied electric voltage as the controlling parameter of the electric drive. So, the generally introduced vector (5.2.2) of the controlling parameters in this case (Figure 5.4) of the electric drive schematization is reduced to one scalar controlling parameter:

$$\mathbf{u} = (u), \quad u = U, \tag{5.5.3}$$

where U is the electric voltage supplied to the direct current electric motor of the considered electric drive.

All the mechanical constituents of the electric drive with the joined working parts (Figure 5.4) can be reduced to the equivalent rotating part with the angular velocity q_1, but all the electrical constituents of the electric drive can be reduced to the equivalent electrical scheme of the direct current electric motor as it is shown in Figure 5.4. So, the generally represented mathematical model (5.2.3) of the processes in the electric drive as the automation object has the view related to the state (5.5.2) and controlling (5.5.3) parameters:

$$\frac{dq_1}{dt} = \frac{1}{J} M(q_1, q_2), \quad \frac{dq_2}{dt} = -\frac{n_G B_e}{L_e} q_1 - \frac{R_e}{L_e} q_2 + \frac{u}{L_e}, \tag{5.5.4}$$

$$q_1(t_0) = q_{1(0)}, \quad q_2(t_0) = q_{2(0)}, \tag{5.5.5}$$

where $M(q_1, q_2)$ is the rotational moment shown in Figure 5.4 as $M(\omega, I)$ and applied to the equivalent rotating part representing the mechanical constituents of the considered electric drive with the working parts; J is the moment of inertia of the equivalent rotating part representing the mechanical constituents of the considered electric drive with the working parts; R_e, L_e and B_e are the resistance, the inductance and the electromechanical parameter of the direct current electric motor; $q_{1(0)}$ and $q_{2(0)}$ are the given values of the angular velocity of the equivalent rotating part representing the mechanical constituents of the considered electric drive with the working parts and the given value of the electric current in the circuit of the winding of the rotor of the direct current motor at the given initial time $t = t_0$.

The mathematical model (5.5.4) and (5.5.5) represents the electromechanical processes in the electric drive, and the view of this model is in full agreement with the general view (2/3) of the model representing the processes in the automation object. Similar models are used in the previous research [34, 35, 41, 42], and it allows to have a lot of interesting results related to the processes in the electric drives.

As it is noted, the mathematical model (5.5.4) and (5.5.5) represents the processes in the electric drive as in the automation object, and to have the mathematical model of this electric drive as the automation object it is necessary to introduce the controlled parameters (5.2.4). In this considered example of the electric drive (Figure 5.4), the angular velocity of the output shaft of the reduction gear defines the operational movements of the working parts, so it is natural to consider this angular velocity as the controlled parameter. So, if the angular velocity of the output shaft of the reduction gear is considered as the controlled parameter of the considered electric drive (Figure 5.4) defined by the state (5.5.2) and controlling (5.5.3) parameters, then the generally introduced vector of controlled parameters (5.2.4) is reduced to the one scalar parameter:

$$\mathbf{x} = (x)\,, \quad x = q_1, \tag{5.5.6}$$

where x is the controlled parameter of the considered electric drive.

It is necessary to highlight that the definition (5.5.6) gives the controlled parameter in the direct view of the dependence on the state parameters (5.5.2) without the generally envisaged differential equations (5.2.5). At the same time, the definition (5.5.6) of the controlled parameter in the considered particular example of the electric drive is the result of the following definition agreed with the general imagination (5.2.5):

$$\frac{dx}{dt} = \frac{dq_1}{dt}, \quad x\,(t_0) = q_{1(0)}. \tag{5.5.7}$$

It is clear that the differential equation (5.5.7) can be integrated analytically, and it leads to the definition (5.5.6). Thus, the mathematical model of the considered electric drive (Figure 5.4) as the automation object is proposed in the view (5.5.4), (5.5.5), and (5.5.7) agreed with the generalized theoretical view (5.2.3) and (5.2.5).

5.5.2 Energy saving controls for transient modes

The transient mode of the automation object is generally defined by the values in the initial conditions presented in the mathematical model (5.2.3) and (5.2.5), as well as by the final condition (5.3.1). The correspondent choosing of such initial and final conditions allows to research any interesting transient mode of the considered automation object.

The considering example of the electric drive is envisaged not to research all possible transient modes in the electric drive, but to illustrate some possible applications of the developed general theory of automation systems with energy and resource-saving functions. So, the considered example of the transient mode must give clear illustrations of the general theory, and the most suitable example for this, is the transient mode of the electric drive (Figure 5.4) related to the acceleration from the state of the rest to the given angular velocity of the output shaft of the reduction gear during the given time. Such transient modes of the considered electric drive (Figure 5.4) correspond to the following initial values (5.5.5) and the following final condition:

$$t_0 = 0, \quad q_{1(0)} = 0, \quad q_{2(0)} = 0, \tag{5.5.8}$$

$$x\left(t_f\right) = \omega_f, \tag{5.5.9}$$

where $x\left(t\right)$ is the function defining the dependence of the controlled parameter of the considered electric drive on the time; t_f and ω_f are the given value of the transient mode time and the required angular velocity of the output shaft of the reduction gear at the final of the transient mode of the considered electric drive.

While only the energy-saving controls are envisaged for the researched transient mode of the considered electric drive, the generally introduced vector (5.3.3) of consumed resources is reduced to the one scalar parameter representing the consumed electric energy:

$$\mathbf{r}\left(t\right) = \left(r\left(t\right)\right), \quad r\left(t\right) = \int_0^t I\left(\tau\right) U\left(\tau\right) d\tau, \tag{5.5.10}$$

where $r\left(t\right)$ is the electric energy already consumed by the considered electric drive before the time moment t; $I\left(t\right)$ and $U\left(t\right)$ are the functions of the time representing the electric current in the winding of the rotor and the electric voltage supplied to the direct current motor of the considered electric drive.

The definitions (5.5.10) and (5.5.2), (5.5.3) allow us to define the objective function required for the formulation of the related optimization problem in the suitable view of the ordinary differential equation with the related initial condition as the complement to the mathematical model of the considered electric drive:

$$\frac{dF}{dt} = q_2 u, \quad F(t_0) = F_0, \tag{5.5.11}$$

where F is the value of the objective function representing the electrical energy consumption of the considered electric drive during the time; F_0 is the amount of the consumed electric energy before the initial time.

The mathematical model (5.5.4), (5.5.5), (5.5.7), and (5.5.11) in the form of the system of the ordinary differential equations with the corresponding initial conditions allows to define the state of the processes, the controlling parameter, and the objective function for the considered electric drive during the time for the given controlling parameter (5.5.3). In agreement with the general approach (5.3.8), the controlling parameter for the considered electric drive (Figure 5.4) is assumed in the following particular view [34, 35]:

$$u(t) = a_0 + a_1 t, \tag{5.5.12}$$

where a_0, a_1 are the parameters defining the related time modes involved to the controlling parameter representing the electric voltage supplied to the direct current electric motor.

Due to the assumption (5.5.12) for the controlling parameter, the mathematical model (5.5.4), (5.5.5), (5.5.7), and (5.5.11) of the considered electric drive (Figure 5.4) allows to define the controlled parameter representing the angular velocity of the output shaft of the reduction gear and the objective function representing the consumed energy as the following relations:

$$x = x(t; a_0, a_1), \tag{5.5.13}$$

$$F = F(t; a_0, a_1). \tag{5.5.14}$$

The relations (5.5.13) and (5.5.14) are the particular views of the generally introduced relations (5.3.12) and (5.3.13) related to the particular example of the considered electric drive (Figure 5.4). So, the generally represented optimization problem (5.3.14)–(5.3.16) for the particular example deals with the defining of the energy saving controls for the researched transient modes of the considered electric drive has the following particular view:

$$F(t_f; a_0, a_1) \to \min, \tag{5.5.15}$$

$$x\left(t_f; a_0, a_1\right) = \omega_f. \tag{5.5.16}$$

In the optimization problem (5.5.15) and (5.5.16), it is assumed that the generally required restriction (5.3.16) is satisfied already for the parameters a_0, a_1 and the time t_f. To solve the optimization problem (5.5.15) and (5.5.16), it is reasonable to represent the equality constraint (5.5.16) in the resolved form, as it is made in the articles [34, 35]:

$$a_1 = f\left(a_0; t_f, \omega_f\right), \tag{5.5.17}$$

where $f\left(a_0; t_f, \omega_f\right)$ is the function, which can be built numerically.

Due to the relation (5.5.17), it is possible to consider the objective function (5.5.15) as the function of one variable:

$$F\left(a_0\right) = F\left(t_f; a_0, f\left(a_0; t_f, \omega_f\right)\right). \tag{5.5.18}$$

Thus, the constrained optimization problem (5.5.15) and (5.5.16) can be reduced to the equivalent unconstrained optimization problem for the objective function (5.5.18):

$$F\left(a_0\right) \to \min. \tag{5.5.19}$$

The unconstrained optimization problem (5.5.19) with the relation (5.5.17) gives the solution of the constrained optimization problem (5.5.15) and (5.5.16), and it allows to define the energy saving controls for the researched transient modes of the considered electric drive (Figure 5.4).

5.5.3 Computer simulations of the energy saving controls

The first differential equation (5.5.4) must be complemented by the definition of the rotational moment $M\left(q_1, q_2\right)$ applied to the equivalent rotating part representing the mechanical constituents of the considered electric drive with the working parts, and the following definition is used:

$$M\left(q_1, q_2\right) = \begin{cases} -Bq_1^2 sign\left(q_1\right) + n_G B_e q_2 - M_{df} sign\left(q_1\right), & q_1 \neq 0, \\ \frac{1}{2}\left(\left|n_G B_e \left|q_2\right| - M_{df}\right| + n_G B_e \left|q_2\right| - M_{df}\right) sign\left(q_2\right), & q_1 = 0, \end{cases} \tag{5.5.20}$$

where B and M_{df} are the generalized parameters of the viscous damping and the generalized moment of the dry friction relatively the output shaft of the reduction gear of the considered electric drive.

To solve the system (5.5.4), (5.5.5), (5.5.7), and (5.5.11) of the ordinary differential equations with the related initial conditions for the controlling parameter (5.5.12) and for the rotational moment defined by the nonlinear relation (5.5.20), it is possible to use only the numerical methods, so the corresponded computer simulations represent the single way to research the energy saving control for the researched transient modes of the considered electric drive. To make such computer simulations, the related program maintenance must be developed, and to do it, it is possible and reasonable to use the Scilab free open-source software [43] envisaged for making scientific computations. This Scilab software provides different opportunities to develop various scenarios for scientific computations, including it has own graphical environment Xcos, allowing it to build the models and make computer simulations [44]. The computer model developed by the Xcos tools representing the mathematical model (5.5.4), (5.5.5), (5.5.7), (5.5.11), (5.5.12), and (5.5.20) is presented in Figure 5.5. In this computer model (Figure 5.5) it is accounted that the differential equation (5.5.7) can be integrated analytically and it can be reduced to the relation (5.5.6). The names of the blocks in this model (Figure 5.5) are in agreement with the notions used in the mathematical model (5.5.4), (5.5.5), (5.5.7), (5.5.11), (5.5.12), and (5.5.20). The not variable numerical data for the computer simulations are assumed in the following view [35]:

$$J = 15\ \mathrm{kg \cdot m^2},\ B = 2,5\ \mathrm{N \cdot m \cdot s},\ M_{df} = 12\ \mathrm{N \cdot m},\ n_G = 10, \quad (5.5.21)$$

$$L_e = 150\ \mathrm{mH},\ R_e = 4,3\ \Omega,\ B_e = 0,82\ \mathrm{N \cdot m \cdot A^{-1}}, \quad (5.5.22)$$

$$t_f = 5\ \mathrm{s}. \quad (5.5.23)$$

The variable data are presented in the discussions of the results of the computer simulations.

The benchmarking of the computer simulations made using the developed model (Figure 5.5) is accomplished for the controlling parameter (5.5.12) in the form of the constant supplied voltage:

$$a_0 \neq 0,\ \ a_1 = 0. \quad (5.5.24)$$

For the controlling parameters (5.12) and (5.24) the angular velocity of the output shaft of the reduction gear and the electric current in the winding of the rotor of the direct current electric motor of the considered electric drive have the following properties:

$$\lim_{t \to \infty} \omega(t) = \omega_\infty,\ \ \lim_{t \to \infty} I(t) = I_\infty, \quad (5.5.25)$$

where ω_∞ and I_∞ are the steady state values.

To define the steady state values from the properties (5.5.25), the equations (5.5.4), (5.5.12), and (5.5.20) are equivalently transformed for the values $\omega_\infty > 0$ and $I_\infty > 0$ to the following view:

$$-B\omega_\infty^2 + n_G B_e I_\infty - M_{df} = 0, \quad -n_G B_e \omega_\infty - R_e I_\infty + a_0 = 0. \tag{5.5.26}$$

Second equality (5.5.26) allows to define the steady state value of the electric current through the steady state value of the angular velocity:

$$I_\infty = \frac{a_0 - n_G B_e \omega_\infty}{R_e}. \tag{5.5.27}$$

The substitution of the electric current in the view (5.5.27) to first equality (5.5.26) leads to the quadratic equation formulated relatively the steady state value of the angular velocity:

$$B\omega_\infty^2 + \frac{n_G^2 B_e^2}{R_e}\omega_\infty + M_{df} - \frac{n_G B_e}{R_e} a_0 = 0. \tag{5.5.28}$$

The quadratic equation (5.5.28) defines steady-state value ω_∞ of the angular velocity, and the relation (5.5.27) defines the steady-state value I_∞ of the electric current. Such defined steady-state values ω_∞ are used to estimate the results of computer simulations, as shown in Figure 5.6 and Figure 5.7. In these figures, we can see that the results of the computer simulations represented by the curves are in agreement with the properties (5.5.25) and with the predicted steady-state values ω_∞ and I_∞ represented by the markers. Such agreement allows to ensure that the results of computer simulations reliably represent the electromechanical processes in the considered electric drive with the direct current motor, and the developed computer model (Figure 5.5) can be used in the optimization procedure to define the energy-saving controls for the researched transient modes of the considered electric drive. The results of the computer simulations for the electric energy consumption of the considered electric drive are presented in Figure 5.8, and they are in agreement with the transient and steady-state modes inherent during the simulation time, as shown in Figure 5.6 and Figure 5.7. Thus, the developed computer model presented in Figure 5.5 allows us to have all the data to define the energy-saving controls for the researched transient modes of the considered electric drive.

The solution of the optimization problem (5.5.15) and (5.5.16) is considered for the values of the parameter a_0 from the interval

$$a_0^{(\min)} \le a_0 \le a_0^{(\max)}, \tag{5.5.29}$$

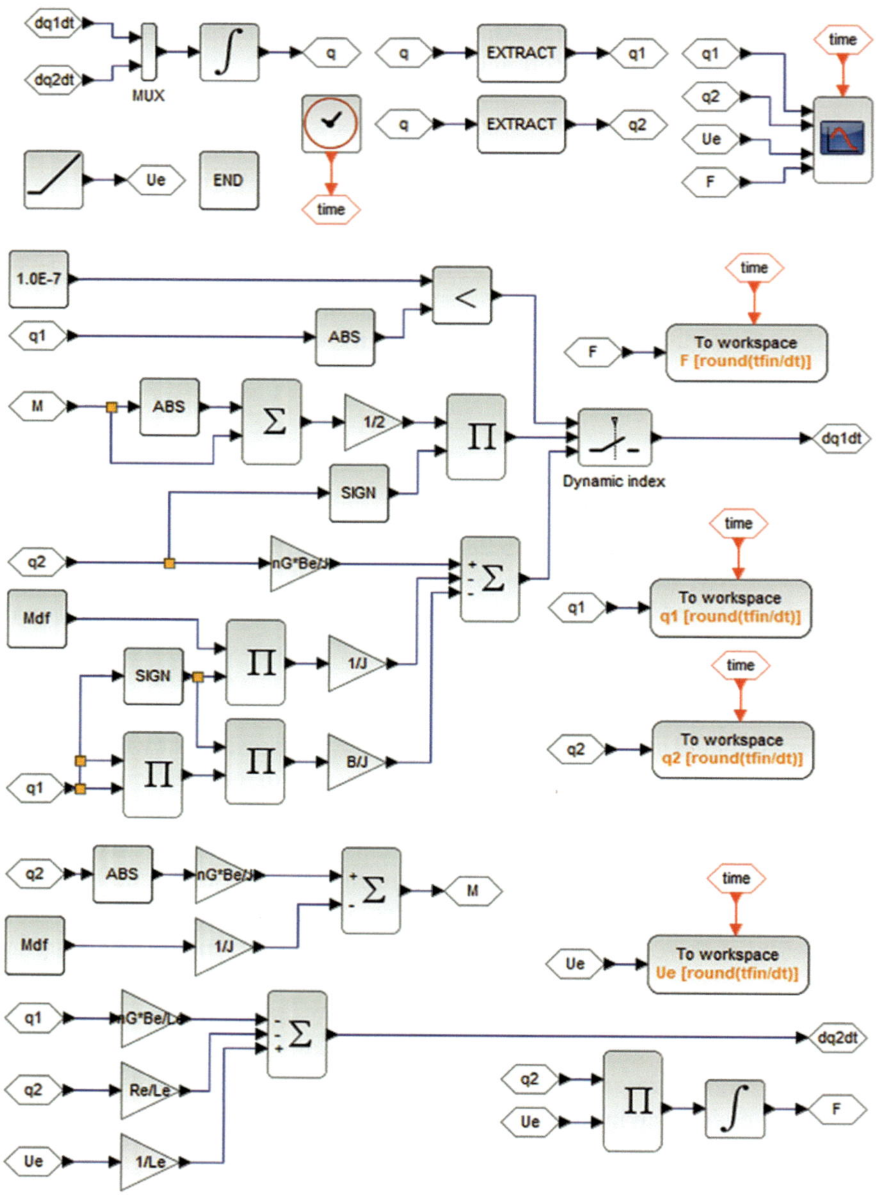

Figure 5.5 Computer model of the electric drive developed by the tools of the Xcos graphical environment envisaged in the Scilab software for scientific computations

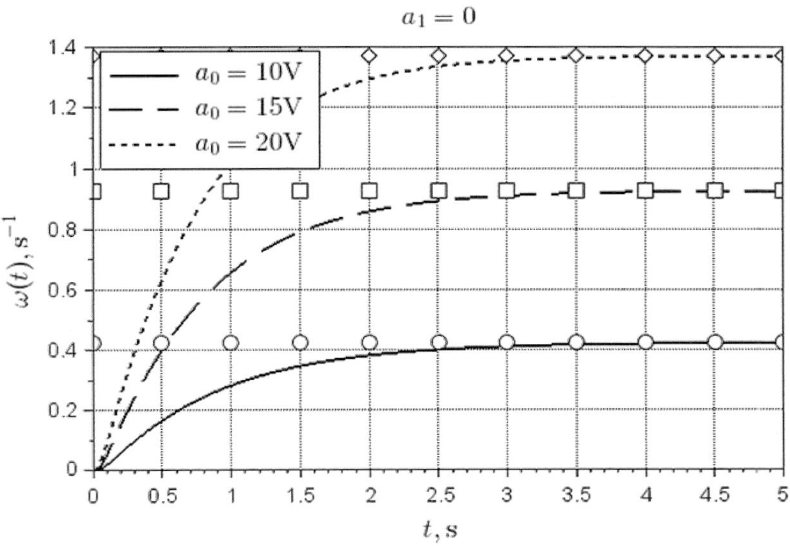

Figure 5.6 Computer simulations (curves) and the related analytical estimations (markers) of the steady-state values of the angular velocity of the output shaft of the reduction gear of the electric drive with the direct current motor

where $a_0^{(\min)}$ and $a_0^{(\max)}$ are the given minimum and maximum values of the parameter a_0.

Instead of the continuous researched interval (5.5.29), the finite set of the values of the parameter a_0 from the researched interval (5.5.29) is considered, and these considered values of the parameter a_0 are defined in the view:

$$\Delta a_0 = \frac{a_0^{(\max)} - a_1^{(\min)}}{N-1}, \quad a_0^{(k)} = a_1^{(\min)} + (k-1)\,\Delta a_0, \quad k = 1, 2, \ldots, N \;, \tag{5.5.30}$$

where N is the count of the considered values, and $a_0^{(k)}$, $k = 1, 2, \ldots, N$ are the considered values of the parameter a_0.

Of course, the exact solution of the optimization problem (5.5.15) and (5.5.16) corresponds to the count $N = \infty$, but it is practically possible to have only the approximate solutions for the different counts $N < \infty$. At the same time, if the count $N < \infty$ is big enough, then the related approximate solutions are close to the exact solution for the optimization problem (5.5.15) and (5.5.16), so such approximate solutions are used further. For each of the values (5.5.30), the equality constraint (5.5.16) can be considered as the

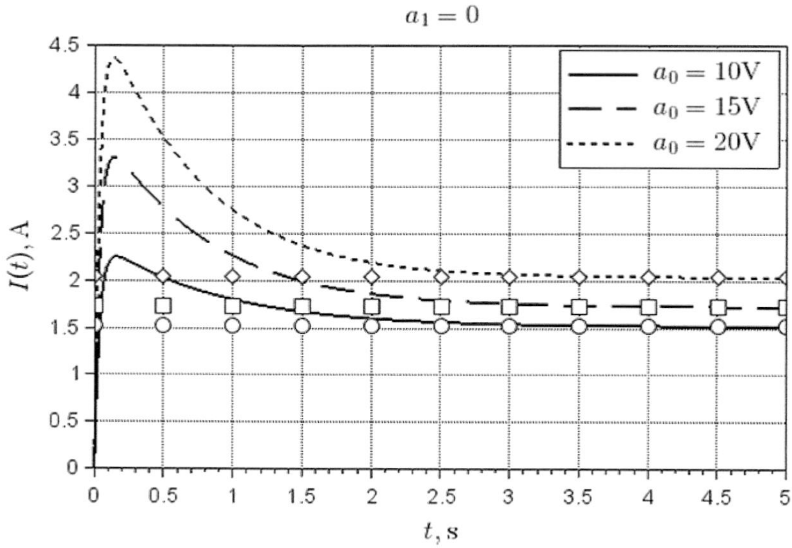

Figure 5.7 Computer simulations (curves) and the related analytical estimations (markers) of the steady-state values of the electric current in the winding of the rotor of the direct current electric motor of the electric drive

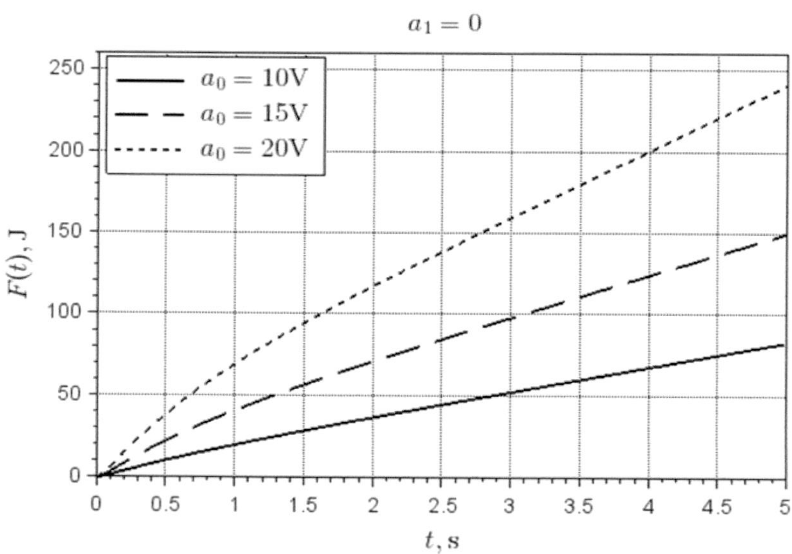

Figure 5.8 Computer simulations of the consumed electric energy for the accelerating electric drive with the direct current motor

nonlinear equation:

$$a_1^{(k)}: \ f\left(t_f; a_0^{(k)}, a_1^{(k)}\right) = 0, \quad k = 1, 2, \ldots, N \ , \tag{5.5.31}$$

where $a_1^{(k)}$, $k = 1, 2, \ldots, N$ are the parameters related to the considered values $a_0^{(k)}$, $k = 1, 2, \ldots, N$, so that the values $a_0 = a_0^{(k)}$ and $a_1 = a_1^{(k)}$ satisfy the equality constraint (5.5.16) for all numbers $k = 1, 2, \ldots, N$.

To solve the equations (5.5.31), it is necessary to use any known numerical method. Such computed values (5.5.30) and (5.5.31) allow to compute the set of the values of the objective function (5.5.14):

$$F_k = F\left(t_f; a_0^{(k)}, a_1^{(k)}\right), \quad k = 1, 2, \ldots, N \ , \tag{5.5.32}$$

where F_k is the value of the objective function computed for the values $a_0 = a_0^{(k)}$ and $a_1 = a_1^{(k)}$ for all numbers $k = 1, 2, \ldots, N$.

The computed values (5.5.32) can be used to define the approximate solution of the optimization problem (5.5.15) and (5.5.16) by means of the searching of the minimum value (5.5.32), and it can be represented in the following view:

$$a_0 = a_0^{(k)}, a_1 = a_1^{(k)}: \ F_{k-1} \leq F_k \leq \ F_{k+1}, \tag{5.5.33}$$

where a_0 and a_1 are the values representing the approximate solution of the constrained optimization problem (5.5.15) and (5.5.16).

The energy saving control for the researched transient mode of the considered electric drive is defined by the relation (5.5.12) with the parameters (5.5.33) representing the solution of the optimization problem (5.5.15) and (5.5.16). To make computer simulations the following date defining the values (5.5.30) are assumed:

$$a_0^{(\min)} = -24\text{V}, \ a_0^{(\max)} = 24\text{V}, \ N = 87 \quad \Rightarrow \quad \Delta a_0 = 0.5\text{V}. \tag{5.5.34}$$

To solve the nonlinear equations (5.5.31) the Interval Halving [45] is used so that the roots are considered inside the preliminary chosen interval with making the fixed count of iterations:

$$-34\text{V}\cdot\text{s}^{-1} \leq a_1^{(k)} \leq 34\text{V}\cdot\text{s}^{-1}, \ i = 15, \ \ k = 1, 2, \ldots, N, \tag{5.5.35}$$

where i is the count of iterations assumed for the Interval Halving.

The proposed scheme for the approximate definition of the energy-saving control for the researched transient modes of the considered electric drive can be easily developed as a scenario of computing in the Scilab software. The results of such computer simulations related to the approximate definitions of the energy-saving controls are presented further. The indirect relations (5.5.17), which are built numerically for the different final values ω_f of the angular velocity of the output shaft of the reduction gear of the electric drive are presented in Figure 5.9. It is shown (Figure 5.9) that the relations (5.5.17) are close to the linear functions even for the nonlinear viscous damping; of course that it is only for the considered, and it cannot be widened on another example without the related justifications. The results of the computer simulations related to the computing of the values of the objective function for their parameters satisfying the equality restriction of the optimization problem (5.5.15) and (5.5.16) are shown in Figure 5.10. It is clearly seen (Figure 5.10) that the objective function has the minimum inside the considered interval (5.5.29) and (5.5.34), and exactly this minimum is related to the energy-saving controls for the researched transient modes of the considered electric drive. It is important that the energy-saving controls are provided by the value $a_0 < 0$ of the parameter defining the constant item of the voltage (5.5.12) supplied to the direct current electric motor of the considered electric drive. So, the results (Figure 5.10) show that the economy of the consumed electric energy for the considered electric drive is about 5–10% compared with the controls related to the value $a_0 = 0$. It is necessary to highlight that the shown economy (Figure 5.10) is principal for the electric drives with a lot of stop-start cycles during the operational time, and the proposed energy-saving automated controls are relevant for implementation in such electric drives.

Although the results presented in Figure 5.9 and Figure 5.10 give a clear imagination of the solutions to the optimization problem with the equality constraint (5.5.15) and (5.5.16), it is important to present also the related energy-saving controls, which are shown in Figure 5.11. It is principally that the energy saving control related to the different final values ω_f of the angular velocity of the output shaft of the reduction gear of the electric drive has close values at the initial time $t = 0$, as seen in Figure 5.11. At the same time, it is not understandable the actual values of the energy-saving controls at the initial time $t = 0$, because the controls presented in Figure 5.11 are built by means of the related approximate solutions of the optimization problem (5.5.15) and (5.5.16). To have more accurate approximate solutions it is necessary to increase the count N of the considered values (5.5.30) to decrease the step Δa_0, but such simulations require a lot of computing, and it

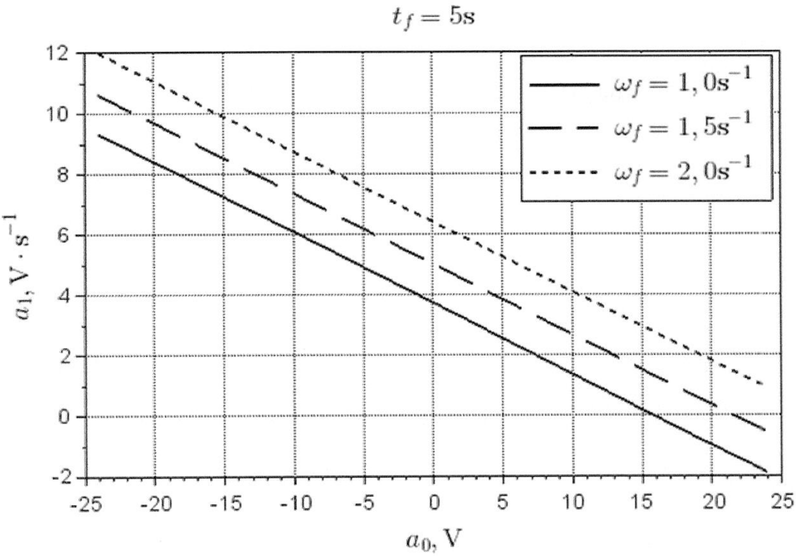

Figure 5.9 Computer simulations of the equality constraints of the optimization problem related to the defining of the energy saving controls.

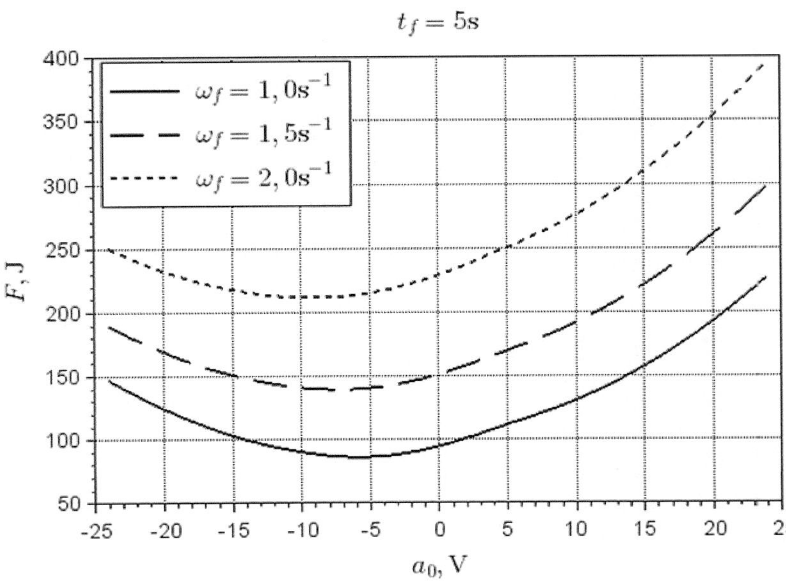

Figure 5.10 Computer simulations of the objective function of the optimization problem related to the defining of the energy-saving controls.

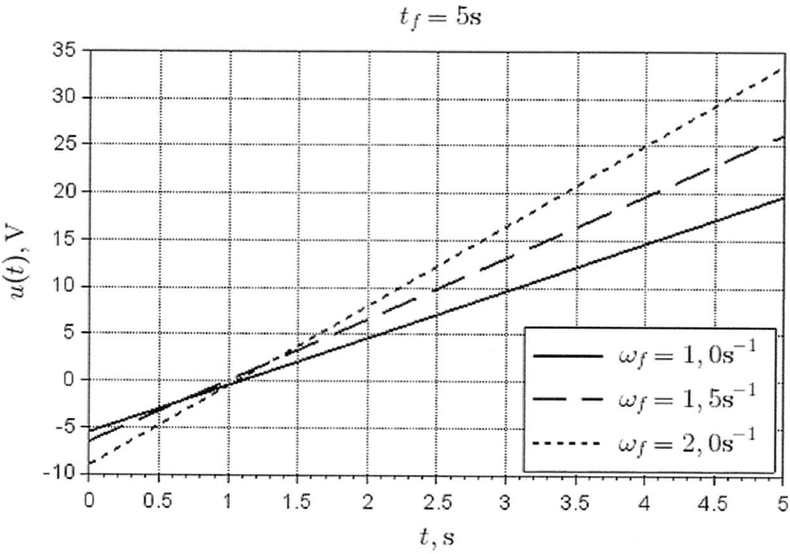

Figure 5.11 Computer simulations of the energy saving controls for the researched transient modes of the considered electric drive.

is not suitable for the Scilab envisaged predominantly for the programming of the computing scenarios, but not for the so big amounts of the computing. The FORTRAN programming language seems more suitable for such computing, but all these are outside this research, and separate attention is required for it. Nevertheless, the shown here results related to the particular example of the electric drive allow us to see the application technique and the possibilities of the developed general theory of energy and resource-saving automation systems for the transient modes of optimal control.

5.6 The PI-Governor with the Water-Saving Function

Water-accumulating vessels are widely used in different water-supplying systems for industrial, agricultural, and householding applications, and it is required to provide the given levels of water in such vessels during the long-term continuous operation so that the automatic PI-governors are often envisaged to do it. The principal purpose of the PI-governor is to provide the required level of water inside the vessel within the given permissible deviation so that the water consumption during the processes of the water level automatic control is not in the primary attention. At the same time, the decreasing of such water consumption is important due to its relevance to sustainable development goals. Besides, it is important also for the thermal and nuclear power plants [46] to use specially purified and prepared water, so the decreasing water consumption allows for minimizing the loading on the water preparation systems, and as a result, reduces the exploitation costs and has cleaner energy production.

5.6.1 PI-Governors for the water level automated control

If it is required to have a stable water level in the vessel (Figure 5.12), the related automatic system is envisaged, and the PI-governors are usually used in such systems. The PI-governor provides the motion of the control valve mounted on the water input pipeline to increase or decrease the flow of the water into the vessel in agreement with the current level of the water measured continuously by means of the level gauge system (Figure 5.12). The mathematical model of the processes in the water accumulating vessels must allow to define the height of the water level, so the generally introduced state parameters are reduced in this considered example to one scalar parameter:

$$\mathbf{q} = (q)\,, \quad q = h, \tag{5.6.1}$$

where h is the height of the water level inside the vessel (Figure 5.12).

The control of the water level is due to the changing of the flow of the water in the water input pipeline, so that the generally introduced controlling parameters (5.2.2) are reduced in this considered example to one scalar parameter (Figure 5.12):

$$\mathbf{u} = (u)\,, \tag{5.6.2}$$

where u is the correction for the flow of the water in the water input pipeline.

The mathematical model of the process of changing the water level inside the vessel has the view (Figure 5.12):

$$\frac{dq}{dt} = \frac{Q_i + u - Q_o}{A_1}, \quad q\,(t_0) = h_0, \tag{5.6.3}$$

where Q_i is the low of the water in the input pipeline; Q_o is the flow of the water in the output pipeline; A_1 is the area of the water mirror in the vessel; t_0 and h_0 are the given initial time and the given height of the water level inside the vessel at the initial time.

The mathematical model (5.6.3) agrees with the generally introduced mathematical model (5.2.3) of the process in automation objects. The area

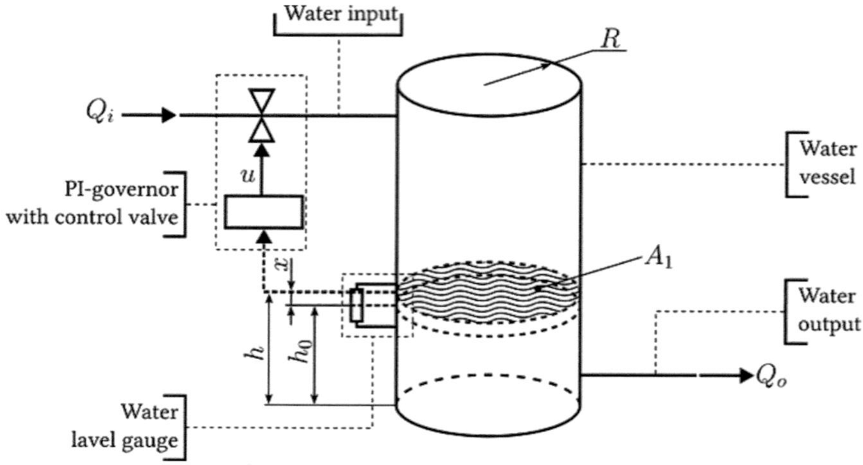

Figure 5.12 The automated control of the water level inside the vessel.

of the water mirror in the vessel with the cylindrical housing can be defined in the view (Figure 5.12):

$$A_1 = \pi R^2, \tag{5.6.4}$$

where R is the internal radius of the vessel.

To have the mathematical model of the water level in the vessel as the automation object, it is necessary to introduce the controlled parameters (5.2.4), which are naturally reduced in his considered example only to one scalar parameter:

$$\mathbf{x} = (x), \quad x = h - h_0, \tag{5.6.5}$$

where x is the difference between the current and initial heights of the water level inside the vessel, as it shown schematically in Figure 5.12.

In the definition (5.6.5) it is assumed that the initial value of the height of the water level is equal to the required height of the water level, which must be provided inside the vessel during the operational time. In this considered example it is possible to have the generally introduced mathematical model (5.2.5) of the automation object in the direct view, and it is suitable to illustrate clearly the general theory of energy and resource-saving automation systems. Indeed, taking into account the notion (5.6.4) of the controlled parameter, the mathematical model (5.6.3) is reduced to the mathematical model of the water level as the automation object:

$$A_1 \frac{dx}{dt} = Q_i + u - Q_o, \quad x(t_0) = x_0, \tag{5.6.6}$$

where x_0 is the initial value of the controlled parameter related with the initial time

It seems that due to the definition (5.6.5) and the initial condition (5.6.3) the initial value of the controlled parameter is predefined $x_0 = 0$, but the values $x_0 \neq 0$ are possible also, and they can be used to simulate the disturbances of the water level inside the vessel. It is necessary to highlight that the mathematical model (5.6.6) is in agreement with the generally introduced linearized mathematical model (5.2.6), but in this considered example the linearized model is built directly from the mathematical model of the processes in the automation object without the parametric identification procedure envisaged in the general case. It provides a clearer illustration of the proposed general theory of energy and resource-saving automation systems because parametric identification is a relatively separate problem requiring particular attention.

If the PI-governor is used to provide automatically the required water level inside the vessel during the operational time, then the controlling parameter (5.6.2) is defined in the following view:

$$u(t) = -K_P x(t) - K_I \int_0^t x(\tau)\, d\tau, \qquad (5.6.7)$$

where K_P and K_I are the parameters of the PI-governor.

The relation (5.6.7) is in agreement with the generally introduced relation (5.4.3), but in this considered example only one controlling parameter and only one controlled parameter are envisaged. To take into account the PI-governor (5.6.7) it is necessary to introduce the new variable:

$$\frac{dy}{dt} = x, \quad y(t_0) = 0. \qquad (5.1)$$

The introduced variable (5.1) is in agreement with the generally introduced variables (5.4.2), but in this considered example only one such variable is required. Due to the introduced variable (5.1), the PI-governor (5.6.7) can be presented in the view:

$$u(t) = -K_P x(t) - K_I y(t). \qquad (5.6.9)$$

So, the mathematical model of the automated system based on the PI-governor for the supporting of the water level inside the vessel involves the relations (5.6.6) complemented by the introduced variable (5.1) and by definition (5.6.9) of the control. This mathematical model allows us to define the controlled parameter (5.6.5) for the given flow $Q_i$of the water in the input pipeline and the given flow Q_o of the water in the output pipeline, as well as for the given values of the K_P and K_I parameters of the PI-governor.

5.6.2 Parameters of the PI-Governor

To define the K_P and K_I parameters of the PI-governor (5.6.7), it is necessary to represent the differential equation (5.6.6) through the introduced variable (5.1) and take into account the control (5.6.9). As the result of the related transformation, the following differential equation is obtained:

$$A_1 \frac{d^2y}{dt^2} + K_P \frac{dy}{dt} + K_I y = Q_i - Q_o. \tag{5.6.10}$$

The differential equation (5.6.10) is in agreement with the generally considered differential equation (5.4.6), but in this considered example only one controlled parameter and only one controlling parameter are envisaged, so only one equation defines the linearized mathematical model. The generally considered characteristic equation (5.4.8) in this considered example related with the differential equation (5.6.10) has the view of quadratic equation:

$$A_1 \lambda^2 + K_P \lambda + K_I = 0, \tag{5.6.11}$$

where λ is the notion for the characteristic root.

The roots of the characteristic equation (5.6.11) have the view:

$$\lambda_{1,2} = \frac{-K_P \pm \sqrt{K_P^2 - 4A_1 K_I}}{2A_1}. \tag{5.6.12}$$

The sense of the parameter A_1 is the area defined by the formula (5.6.4) or other similar, so this parameter can be positive only:

$$A_1 > 0. \tag{5.6.13}$$

Due to the inequality (5.6.13), the characteristic roots (5.6.12) have the negative real parts for all the parameters of the PI-governor satisfying the conditions:

$$K_P > 0, \quad K_I > 0. \tag{5.6.14}$$

The inequalities (5.6.14) defining the possible parameters of the PI-governor are the particular case of the generally formulated conditions (5.4.11). It is necessary to highlight that the characteristic roots (5.6.12) are related with three principal cases corresponding to the real roots, real multiple roots, and complex conjugated roots. To consider these cases, it is suitable to represent the parameters (5.6.14) as

$$K_P = 2K\sqrt{A_1 K_I}, \tag{5.6.15}$$

where $K > 0$ is some numerical parameter.

The relation (5.6.15) is suitable because the variation of the parameter K allows to consider the cases of the characteristic roots (5.6.12), so that the complex conjugated roots correspond to the parameter

$$0 < K < 1. \tag{5.6.16}$$

The real multiple roots corresponded to the parameter

$$K = 1. \tag{5.6.17}$$

Finally, the real roots are corresponded to the parameter

$$K > 1. \tag{5.6.18}$$

So, the parameters K_P and K_I (5.6.14) can be defined through the parameters $K > 0$ and $K_I > 0$ with the usage of the relation (5.6.15).

The resource efficiency of the automated system for supporting the required water level is reduced in this particular example to the minimization of the water consumption during the operational time of the considered water accumulating vessel (Figure 5.12). So, the generally introduced vector (5.3.4) of amounts of different types of resources already consumed during the operation of the considered automation object is reduced to the one scalar parameter defining the water consumption:

$$\mathbf{r} = (r), \quad r(t) = \int_0^t (Q_i(\tau) - u(\tau))\, d\tau, \tag{5.6.19}$$

where r is the amount of consumed water during the operational time of the considered water-accumulating vessel (Figure 5.12).

The differential equation to define the consumed water (5.6.19) has the view:

$$\frac{dr}{dt} = Q_i - u, \quad r(t_0) = r_0, \tag{5.6.20}$$

where r_0 is the given amount of the consumed water before the given initial time $t = t_0$.

It is necessary to highlight that the relations (5.6.20) are in agreement with the generally introduced relations (5.3.5), but this particular example is simpler.

To consider the water saving PI-governor, it is necessary to introduce the following value:

$$\frac{dF}{dt} = -u, \quad F(t_0) = 0, \tag{5.6.21}$$

where F is the amount of the consumed water required to support the necessary level of the water inside the vessel (Figure 5.12).

The parameter (5.6.21) can be computed by using the mathematical model (5.6.6), (5.1), (5.6.9), (5.6.15), and (5.6.21) and the suitable numerical method to solve the related system of the ordinary differential equations with the initial conditions. As a result , the following relation can be built:

$$F = F(t; K, K_I). \tag{5.6.22}$$

The relation (5.6.22) allows formulate the objective function:

$$F(K, K_I) = \max_{t_0 \le t \le t_f} F(t; K, K_I), \tag{5.6.23}$$

where t_f is the given final time.

The objective function (5.6.23) allows to formulate the optimization problem related to the defining of the PI-governor with minimal water consumption has the following view:

$$F(K, K_I) \to \min, \quad K > 0, K_I > 0. \tag{5.6.24}$$

The objective function (5.6.23) also allows to formulate the optimization problem related to the defining of the PI-governor with the maximal water consumption has the following view:

$$F(K, K_I) \to \max, \quad K > 0, K_I > 0. \tag{5.6.25}$$

The comparison of the PI-governors defined by means of the optimization problems (5.6.24) and (5.6.25) allows to see the effect of the implementation of the resource-saving functions related to the decreasing of the water consumption during the operational time of the considered water accumulating vessel (Figure 5.12).

To solve numerically the optimization problems (5.6.24) and (5.6.25), the following values of the parameters $K > 0$ and $K_I > 0$ are considered:

$$K^{(\min)} \le K \le K^{(\max)}, \quad K_I^{(\min)} \le K_I \le K_I^{(\max)}. \tag{5.6.26}$$

where $K^{(\min)} > 0$, $K^{(\max)} > 0$ and $K_I^{(\min)} > 0$, $K_I^{(\max)} > 0$ are the given minimum and maximum values assumed for the parameters $K > 0$ and $K_I > 0$.

Instead of the continuous intervals (5.6.26), only the finite sets of the parameters K and K_I are considered:

$$\Delta K = \frac{K^{(\max)} - K^{(\min)}}{N_K - 1}, \quad K^{(i)} = K^{(\min)} + (i-1)\,\Delta K, \quad i = 1, 2, \ldots, N_K, \tag{5.6.27}$$

$$\Delta K_I = \frac{K_I^{(\max)} - K_I^{(\min)}}{N_I - 1}, \quad K_I^{(i)} = K_I^{(\min)} + (i-1)\,\Delta K_I, \quad i = 1, 2, \ldots, N_I, \tag{5.6.28}$$

where N_K and N_I are the counts of the considered values for the parameters K and K_I; $K^{(i)}$, $i = 1, 2, \ldots, N_K$ and $K_I^{(i)}$, $i = 1, 2, \ldots, N_I$ are the considered values of the parameters K and K_I.

Due to the relation (5.6.23) and the considered values (5.6.27) and (5.6.28) of the parameters K and K_I it is possible to define the related values of the objective function:

$$F_{ij} = F\left(K^{i}, K_I^{(j)}\right), \quad i = 1, 2, \ldots, N_K, \quad j = 1, 2, \ldots, N_I, \tag{5.6.29}$$

where F_{ij} are the values of the objective function related to the considered parameters $K^{(i)}, i = 1, 2, \ldots, N_K$ and $K_I^{(j)}, j = 1, 2, \ldots, N_I$.

The values (5.6.27)–(5.6.29) allow to represent the approximate solutions of the optimization problems (5.6.24) and (5.6.25):

$$\breve{F} = F_{ij} : \quad F_{ij} \leq F_{kl} \;\; \forall k \neq i \wedge l \neq j, \quad \breve{K} = K^{(i)}, \; \breve{K}_I = K_I^{(i)}, \tag{5.6.30}$$

$$\frown{F} = F_{ij} : \quad F_{ij} \geq F_{kl} \;\; \forall k \neq i \wedge l \neq j, \quad \frown{K} = K^{(i)}, \; \frown{K}_I = K_I^{(i)}, \tag{5.6.31}$$

where $\breve{F}$ is the value of the objective function and $\breve{K}$, $\breve{K}_I$ are the values of the parameters K and K_I representing the solution of the minimization problem (5.6.24); $\frown{F}$ is the value of the objective function and $\frown{K}$, $\frown{K}_I$ are the values of the parameters K and K_I representing the solution of the maximization problem (5.6.25).

The ratio between the values $\breve{F}$ and $\frown{F}$ allows to estimate the effect of the implementation of the water-saving functions for the PI-governor of the water level during the operational time of the water-accumulating vessel (Figure 5.12). It is necessary to highlight that the solutions (5.6.30) and

(5.6.31) of the optimization problems (5.6.24) and (5.6.25) are approximate, because of the counts of the considered values (5.6.27) and (5.6.28) are finite, so that $N_K < \infty$ and $N_I < \infty$. At the same time, these approximate solutions (5.6.30) and (5.6.31) are close to the exact solution, if the counts N_K and N_I of the considered values (5.6.27), (5.6.28) are big enough. Besides, it is not necessary to make the approximate solutions (5.6.30) and (5.6.31) so close to the exact solutions for the illustration of the proposed general theory of automation systems with energy and resource-saving functions. If it is required to have the approximate solutions (5.6.30) and (5.6.31) very close to the exact solutions of the considered optimization problems (5.6.24) and (5.6.25), then the programming languages envisaged to develop the scenarios of the computing are not suitable due to the slow computing, and the FORTRAN programming language is the most suitable in such cases. Nevertheless, it is possible to make the computer simulations required to find the approximate solutions (5.6.30) and (5.6.31) by means of different computer technologies.

5.6.3 Computer simulations of the water saving PI-Governor

Due to the significant amount of required computing, it is difficult to define the water saving PI-governor without the numerical methods and computer simulations even for the linear mathematical models (5.6.6), (5.1), (5.6.9), (5.6.15), (5.6.20), and (5.6.21) and to solve the optimization problems (5.6.24) and (5.6.25) in the approximate view (5.6.30) and (5.6.31). So, the Scilab free open-source software [43, 44] is used to make the required computer simulations, because the proposed solution scheme can be represented as the scenario of the typical computing.

Figure 5.13 presents the computer model of the processes in the water-accumulating vessel (Figure 5.12) based on the mathematical models (5.6.6), (5.1), (5.6.9), (5.6.15), (5.6.20), and (5.6.21) and developed by tools of the Xcos graphical environment envisaged in the Scilab. This developed computer model (Figure 5.13) assumes the possibility of the research about the compensations by means of the PI-governor of the initial disturbance and random disturbances during the given operational time of the water accumulating vessels. Besides, the computer model (Figure 5.13) is used to build the water-saving PI-governors. The following values of the not variable input data are used for making computer simulations:

$$R = 1.75\,\mathrm{m}, \quad Q_i = 0.125\mathrm{m}^3 \cdot \mathrm{s}^{-1}, \quad t_0 = 0, \quad t_f = 300\ \mathrm{s},$$

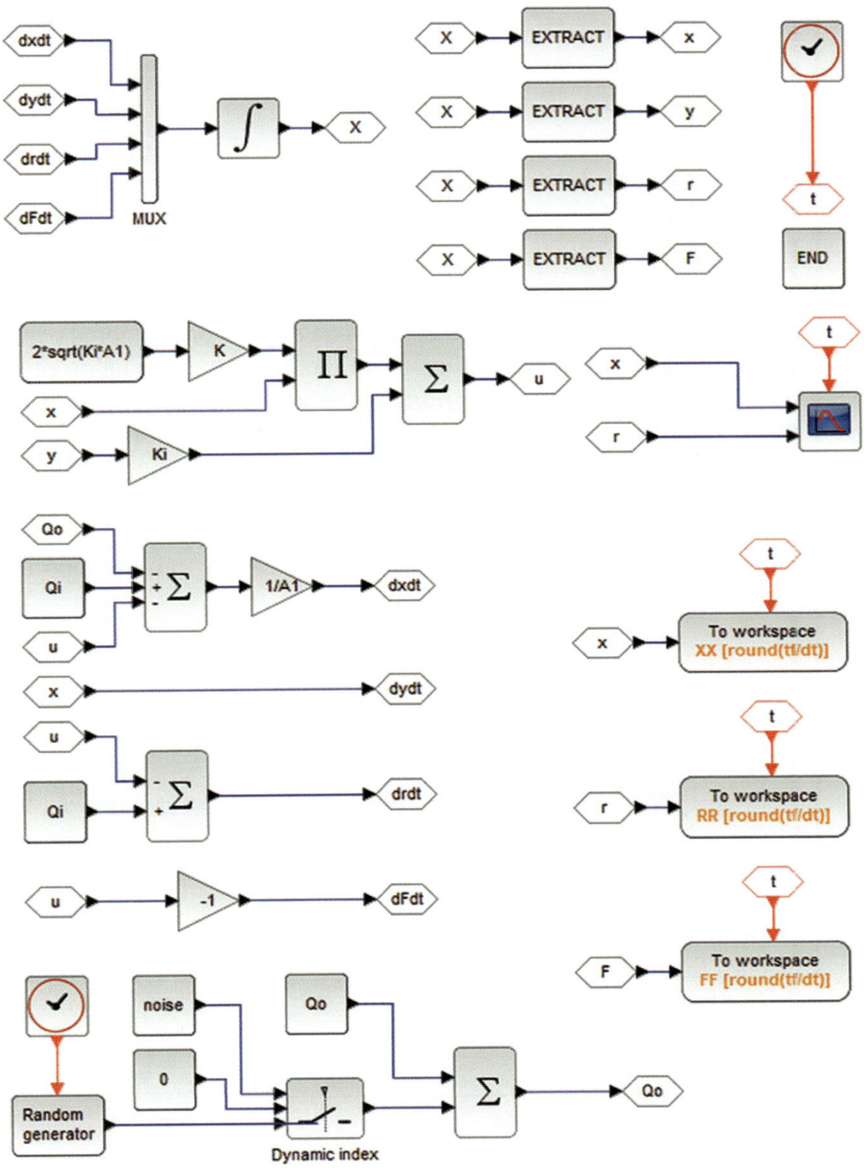

Figure 5.13 Computer model of the water accumulating vessel developed by the tools of the Xcos graphical environment envisaged in the Scilab software for scientific computations.

$$K^{(\min)} = 0.5, \quad K^{(\max)} = 1.5, \quad K_I^{(\min)} = 0.01 \text{ m}^2 \cdot \text{s}^{-2}, \\ K_I^{(\max)} = 0.05 \text{ m}^2 \cdot \text{s}^{-2}. \tag{5.6.32}$$

The variated and specific values of the required input data are introduced further during the related results discussions.

To research the influence of the parameters K and K_I defining the PI-governor, it is considered the transient mode related to the compensation of the initial disturbance for the following values of the input data:

$$x_0 = -0.1 \text{ m}, \quad Q_o = 0.125 \text{m}^3 \cdot \text{s}^{-1}. \tag{5.6.33}$$

The results of the computer simulations prove that the proportional part of the PI-governor has a significant influence on the maximum deviation of the level (Figure 5.14) and on the controlling water consumption (Figure 5.15), although the water consumption effects are not so observable on the general water consumption (Figure 5.16). Nevertheless, the computer simulation results show that it is possible to develop the PI-governors providing minimal water consumption peaks during the time of the transient process of the compensation of the water level initial disturbance, which can be important

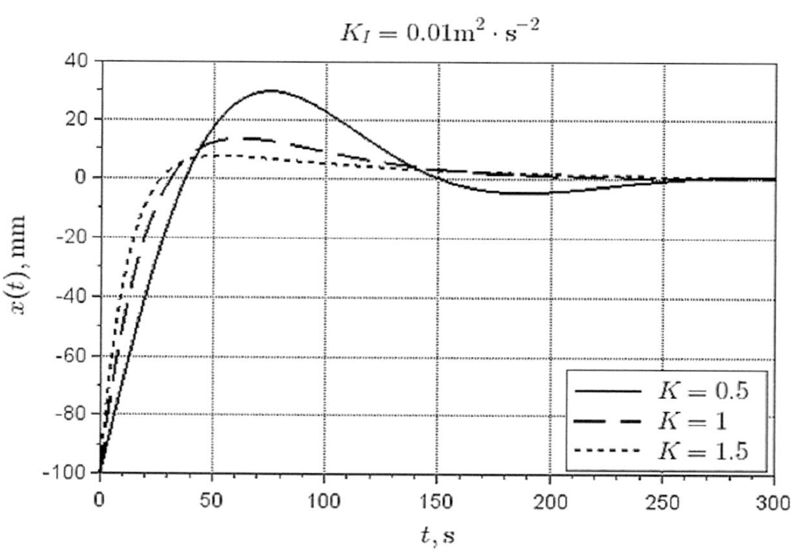

Figure 5.14 Computer simulations of the water level inside the vessel during the compensation of the initial disturbance for different parameters defining the proportional part and for the given value of the parameter defining the integral part of the PI-governor.

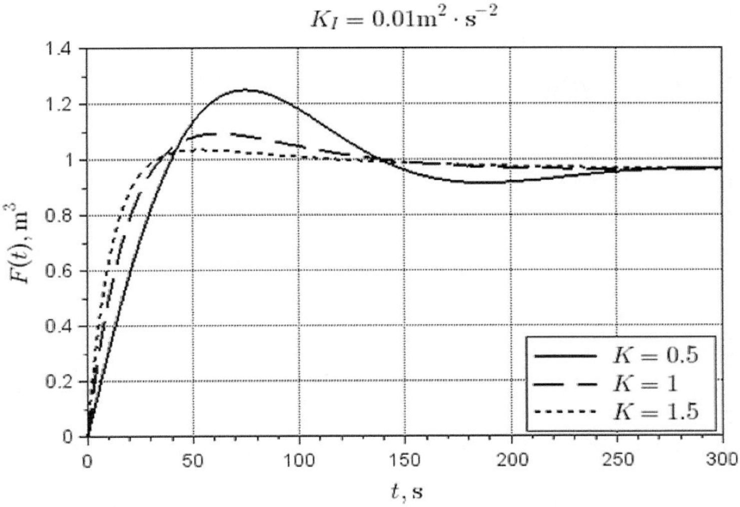

Figure 5.15 Computer simulations of the controlling water consumption during the compensation of the initial disturbance of the water level inside the vessel for different parameters defining the proportional part for the given value of the parameter defining an integral part of the PI-governor.

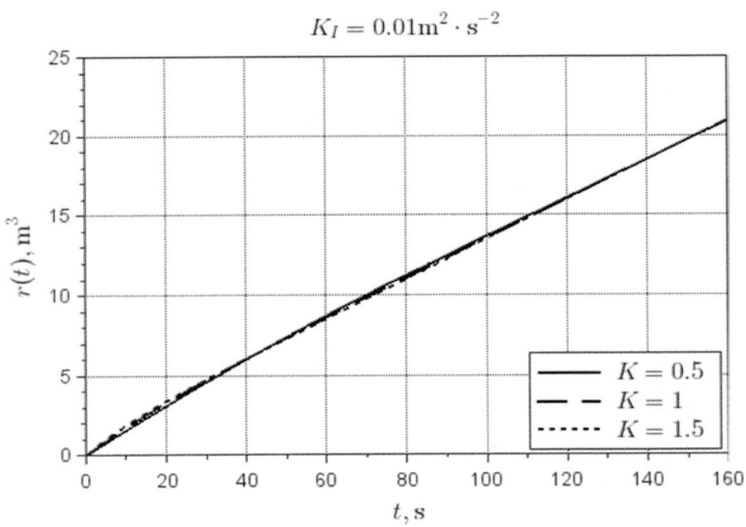

Figure 5.16 Computer simulations of the total water consumption during the compensation of the initial disturbance of the water level inside the vessel for different parameters defining the proportional part with the given value of the parameter defining an integral part of the PI-governor.

to stabilize the water flows to cover the changing water consumption in the complicated water supply systems.

The results of the computer simulations prove that the parameter defining an integral part of the PI-governor has no significant influence on the maximum deviation of the level (Figure 5.17) and on controlling water consumption (Figure 5.18). The influences of the integral part of the PI-governor are noticeable in the duration of the transient process of the compensation of the initial disturbance of the water level height inside the vessels, as illustrated in Figure 5.17 and Figure 5.18, although the water consumption effects are not so observable on the general water consumption (Figure 5.19). Nevertheless, the time of the transient processes related to the compensation of the initial disturbances of the water level inside the vessels can be important for some classes of automation systems. So, the time of the transient processes is the principal characteristic of the PI-governor, and it must be accounted for to design suitable PI-governors with the minimum peaks of controlling water consumption during the inherent transient processes. Due to these circumstances, it is possible to choose the PI-governor providing the minimum peaks of the controlling water consumption for the required limited time of the transient processes by corresponding selection of the proportional

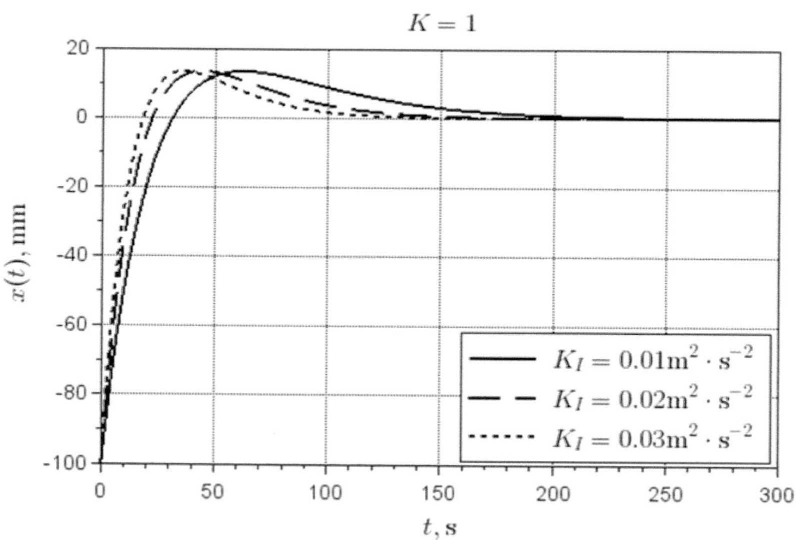

Figure 5.17 Computer simulations of the water level inside the vessel during the compensation of the initial disturbance for different parameters defining the integral part and for the given value of the parameter defining the proportional part of the PI-governor.

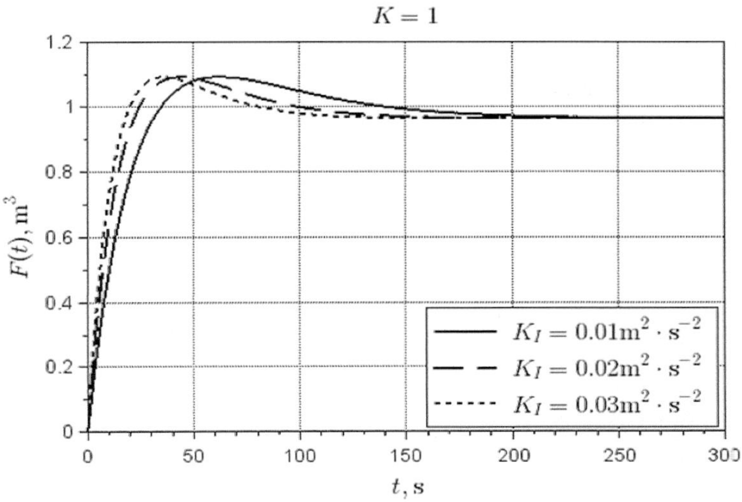

Figure 5.18 Computer simulations of the controlling water consumption during the compensation of the initial disturbance of the water level inside the vessel for different parameters defining the integral part and for the given value of the parameter defining the proportional part of the PI-governor.

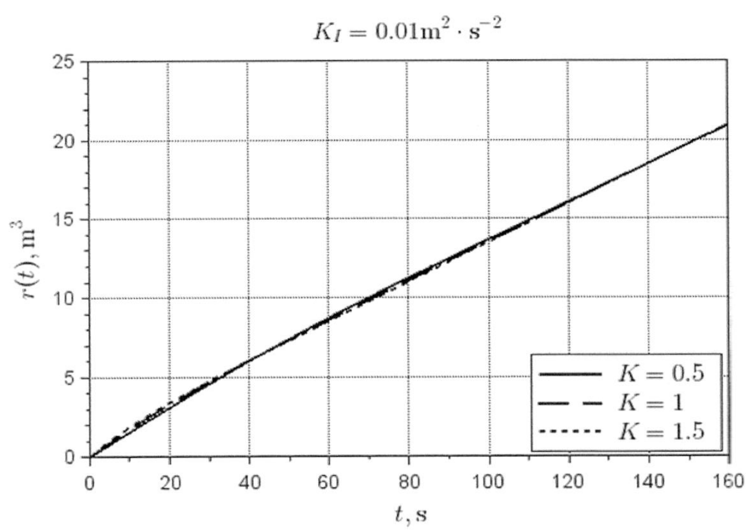

Figure 5.19 Computer simulations of the total water consumption during the compensation of the initial disturbance of the water level inside the vessel for different parameters of the integral part and for the given value of the parameter defining the proportional part of the PI-governor.

part for this PI-governor. It is necessary to highlight that it is impossible to decrease the water consumption by implementation of the PI-governors with the water saving function because it is required to provide the water consumption required for the consumers, but it is possible to decrease the peaks of the water consumption during the operation of the water supply system.

The calculations and the computer simulations lead to the approximate solutions (5.6.30) and (5.6.31) in the following view:

$$\breve{F} = 1.0347997 \text{ J}, \;\; \breve{K} = 1.5, \quad \breve{K}_I = 0.05 \text{ m}^2 \cdot \text{s}^{-2}, \tag{5.6.34}$$

$$\frown{F} = 1.2491304 \text{ J}, \;\; \frown{K} = 0.5, \;\; \frown{K}_I = 0.01 \text{ m}^2 \cdot \text{s}^{-2}. \tag{5.6.35}$$

The comparison of the values $\breve{F}$ and $\frown{F}$ (5.6.34) and (5.6.35) proves that it is possible to decrease the peak values of the controlling water consumption required to compensate for the initial disturbance of the water level in the vessel by related choosing of the parameters of the PI-governor. Some results of computer simulations on the processes related to the water level automatic control in the accumulation vessel for the minimum consumption PI-governor defined by the parameters (5.6.34) and the maximum consumption PI-governor defined by the parameters (5.6.35) are presented in Figure 5.20 and Figure 5.21. It is important, that in this considered example, the minimum

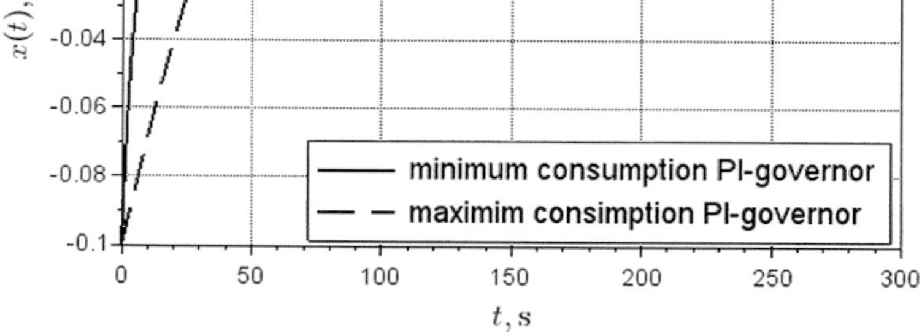

Figure 5.20 Computer simulations of the water level inside the vessel during the process of the compensation of the initial disturbance for the PI-governors with the parameters providing the minimum and maximum peaks of the controlling water consumption.

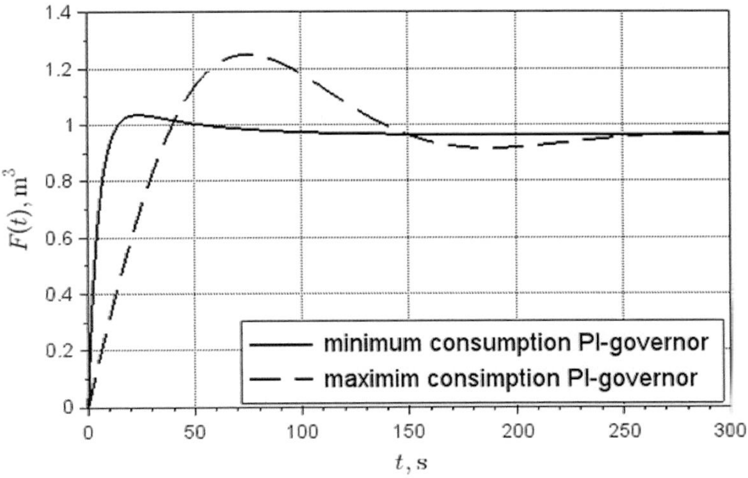

Figure 5.21 Computer simulations of the controlling water consumption during the process of the compensation of the initial disturbance of the water level inside the vessel for the PI-governors with the parameters providing the minimum and maximum peaks of the controlling water consumption.

consumption PI-governor provides the minimum time for the transient mode to compensate for the initial disturbance of the water level in the vessel. So, it is proved that it is principally possible to design the PI-governors for the water level automatic control providing minimal peaks of the controlling water consumption and the minimal duration of the transient modes. It is important for thermal and nuclear power plants [46] to use purified and prepared water.

The parameters (5.6.34) and (5.6.35) are defined to decrease the maximum peaks of the controlling water consumption to compensate for the initial disturbance of the water level in the vessel during the related transient modes. It is important to compare the PI-governors with the parameters (5.6.34) and (5.6.35) on the steady-state modes related to the given average supply Q_i of the water inside the vessel, but with the random consumption of the water outside the vessel, because such mode is typical for the water-supplying systems. To make the related computer simulations, instead of the value (5.6.33), the following input data are assumed:

$$x_0 = 0, \quad Q_o = \bar{Q}_o + \tilde{Q}_o(t)\,, \tag{5.6.36}$$

where $\bar{Q}_o$ is the given value of the average consumption of the water; $\tilde{Q}_o(t)$ is the function of the time defining the random disturbances of the water consumption.

The computer simulations are made for the water consumption (5.6.36) with the following parameters:

$$\bar{Q}_o = 0.125\text{m}^3 \cdot \text{s}^{-1}, \quad \left|\tilde{Q}_o(t)\right| \leq \bar{Q}_o/10. \tag{5.6.37}$$

The parameters (5.6.38) are correspondent to the case of the equal average values of the water supplying defined in the data (5.6.32) and the water consumption; the random disturbances of the water consumption are assumed as not bigger than 10% of the average water consumption, and such disturbances are 5 seconds each. Simulations of such disturbances are envisaged in the developed computer model (Figure 5.13), and some results of such computer simulations are presented in Figure 5.22 and Figure 5.23. It is shown that the minimum consumption PI-governor defined by the parameters (5.6.34) provides the minimum controlling water consumption peaks then the maximum consumption PI-governor defined by the parameters (5.6.35) even for the steady-state mode with the random disturbances of the total water usage by the consumers (Figure 5.23). Besides, the minimum consumption PI-governor provides the water level changing within smaller intervals than the maximum consumption PI-governor (Figure 5.22). It proves that the water-saving functions are in agreement with the principal purposes of the

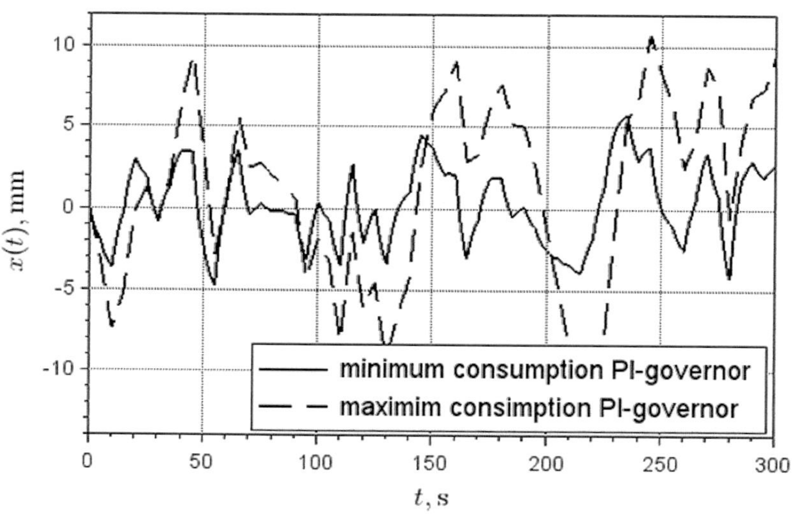

Figure 5.22 Computer simulations of the water level inside the vessel under the random disturbances of the total water consumption for the PI-governors with the parameters providing the minimum and maximum peaks of the controlling water consumption.

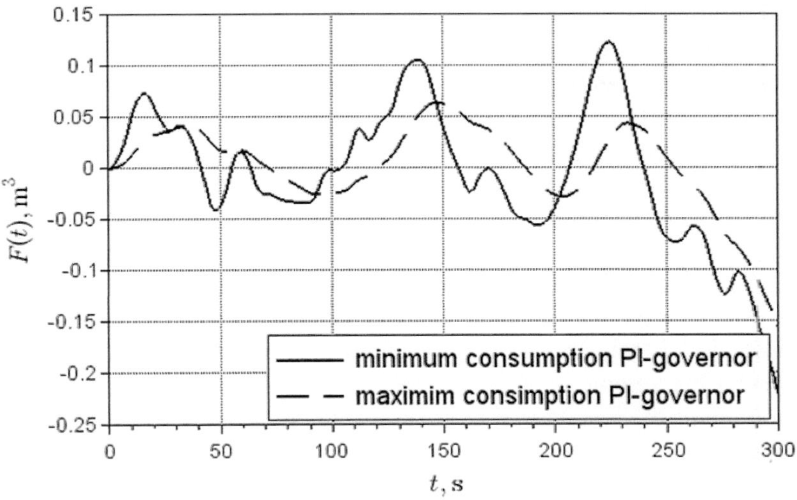

Figure 5.23 Computer simulations of the controlling water consumption under the random disturbances of the total water consumption for the PI-governors with the parameters providing the minimum and maximum peaks of controlling water consumption.

PI-governor in supplying water level within the required interval inside the vessels.

The considered example about the PI-governor for the water level support during the operation of the water accumulating vessels gives a clear imagination about the application technique and about the possibilities of the developed general theory of automated system with energy- and resource-saving functions. In this particular example, it is proved that the resource-saving functions can be in agreement with the main purposes of the automated systems, and moreover, such resource-saving functions can improve the general characteristics of the developed automatic governors. At the same time, the existing experience of the usage of the proposed theory is limited, and it is necessary to make the related research.

5.7 Conclusion

To conclude this research, it is possible to claim that one variant of the general theory of automation systems with energy- and resource-saving functions has been developed. The developed theory is important, because it gives a clear way to develop automated systems with energy and resource-saving functions for different applications, including the Internet of Things. The results allow to formulate the important particular inferences of this research complementing the general claims.

The developed general theory of automation systems with energy- and resource-saving functions includes the formalized approaches providing the mathematical modeling of automation objects. These formalized approaches are based on the mathematical modeling of the processes inherent in the automation object and on the usage of the controlling and controlled parameters to define the state of the automation objects. It is proposed to consider the mathematical model of the processes inherent in the automation object as the digital twins for parametric identification to build the related linearized mathematical models required to design the automatic governors. It is proposed to consider the velocities of energy and resource consumptions related to the exploitation as the results of the processes inherent in the automation objects so as to have the complementary ordinary differential equations with the initial conditions defining the energy and resource consumptions. It is shown that in the general case, the mathematical model of an automation object can be represented by means of the set of the system of the ordinary differential equations with the related initial conditions, so that first of these systems must represent the processes in the automation object with the introduced state and controlling parameters, second of these systems must define the controlled parameters, and third of these systems must define the energy and resource consumptions during the operation of the automation object.

It is proposed to use the optimal control theory with the related optimality criteria to consider energy- and resource-saving operations of automation subjects on the transient modes. The indirect view of the differential equations defining the relation between the controlled and controlling parameters in the general case significantly complicates the consideration of the optimal control problems, so that only the numerical methods are possible. It is proposed to reduce the optimal control problem regarding energy and resource savings for the transient modes of automation objects to related constrained optimization problems, and it is shown that in the general case, such optimization problem is defined by the indirect objective function and the indirect equality and inequality constraints. This objective function as well as these equality and inequality constraints are defined through the set of the scalar parameters representing the weights of the possible controlling modes, which are the given functions of the time forming the control of the automation object. Such representation of the control through the finite controlling modes limits the class of the considered controls, but it is in agreement with the applications. The objective function of such an optimization problem is related to the consumed energy and resources, but the constraints are related to the limited controls and the given initial and final states of the researched transient mode. The principal difficulty of the development of the energy and resource-saving optimal controls for transient modes of automation object is due to the indirectly defined objective functions and the related restrictions of the equivalent optimization problem, so to calculate the values of this objective function and these constraints, it is necessary to solve numerically the non-linear initial-value problem. It leads to a significant amount of computing, so to solve numerically such optimization problems, it is necessary to use the corresponding programming languages, like FORTRAN, and the popular programming languages envisaged to develop the computing scenarios can be used only for not difficult problems or if the low items discretization is used. In any case, such formulated optimization problems related to the automation systems having energy- and resource-saving controls for the transient modes have not been researched yet, so the corresponding researches are required further.

The consideration of the automation systems with the energy- and resource-saving functions on the steady-state modes is reduced to the defining of the parameters of the PID-governors, which are the most used for the automation objects operated under such operational modes. It is shown that the PID-governor influences on the automation object can be represented by means of the complementary first-order differential equation with the initial condition to define an integral part of the regulation law of this PID-governor, and the relation defining the controlling parameters taking into account the proportional, integral, and differential items envisaged in the regulation law of this PID-governor. Only the PI-governor can be accounted for without the complicated transformations of the differential equations of the mathematical model representing the processes in the considered automation object in the general case, but the consideration of the PID-governor requires the related transformations of these differential equations in the general case, although for the linear systems such transformations are simple. It is proposed to reduce the development of the automatic PID-governors with energy- and resource-saving functions to the consideration of the equivalent optimization problem with the objective function and the constraints defined through the parameters of this PID-governor, and it is shown that in the general case, such optimization problem has the inequality constraints only. The objective function of such formulated optimization problems is related to the consumed energy and the resources during the given operational time of the considered automation object with the researched PID-governor, but the constraints of such optimization problems are related to the stability conditions of the corresponding linearized system with the researched PID-governor. The objective function and the constraints of such optimization problem in the general case are defined indirectly so that to compute them for the given value of the parameters of the PID-governor, it is necessary to solve numerically the nonlinear initial value problem and to solve the set of the eigenvalue problem or the set of the high-powered algebraic equations. So, to solve such a formulated optimization problem in the general case, it is necessary to use suitable numerical methods because of the significant amount of the required computations. Such formulated optimization problems related to the automation systems with energy- and resource-saving automated controls for steady-state modes have not been researched yet, so correspondent researches are required further.

The proposed general theory allows the development of energy-saving controls for the transient modes, as shown in the related example about the electric drive with the direct current motors. It is shown that the energy-saving controls, providing the minimum of the electric energy consumption to accelerate the electric drive from the state of the rest to the given angular velocity during the given time, are supplied voltages with the linear time dependences starting from negative voltage related to the motion of the reverse direction of the electric drive. This result is considered in different previous research, and this research is the illustration for the case of quadratic viscous damping. At the same time, it is necessary to research more carefully the value of the negative voltage required at the initial time to provide the energy-saving controls for the acceleration transient modes of the electric drives, because the computed values are different, but close, so these different can be the result of the used discretization to solve numerically the related optimization problem. It is necessary to research further the relation between the value of this negative voltage required at the initial time to provide the energy-saving controls and the value of the moment of dry friction presented in the considered electric drive. In any case, the considered example and the results of the computer simulations approve the agreement of the general theory with the applications about the energy saving controls for the transient modes of the electric drives with the direct current motors.

The considered example of the PI-governor for the automatic control of the water level inside the accumulating vessel allows us to illustrate that the water-saving functions are possible for these applications. At the same time, it is shown that it is impossible to reduce the water consumption by means of the suitable PI-governor, because it is not allowed to restrict the consumers, but the relevant PI-governors permit to decrease the peaks inherent for controlling water consumption required for the compensation of the initial disturbances of the water level inside the water accumulating vessels. Of course, it not only provides the economy of the water resources, but it allows for a decrease in the loading on the water supply systems, and it is important for thermal and nuclear power plants to use the purified and specially prepared water. In this example, it is shown that the resource-saving functions can be in agreement with the main purpose of the automatic governors. Indeed, it is shown that the PI-governor with the minimum peaks of the controlling water consumption provides the water level within the least interval compared with the ordinary PI-governor. It is also shown that the minimum peaks of the controlling water consumption are related to the real roots of the correspondent characteristic equation defining the stability conditions to choose the parameters of the PI-governor. Nevertheless, the principal difficulties of the usage of such an approach are related to the linearization of the differential equations of the mathematical model representing the considered automation object, so it is necessary to develop further suitable approaches for the parametric identification in the cases of the higher order linear ordinary differential equations representing the complicated automation objects.

Acknowledgments

The authors thank Irina Kolupaieva, the colleague, the professor of the Department of Computer-Integrated Technologies, Automation and Robotics at Kharkiv National University of Radio Electronics (Ukraine) for the collaborations in the fields of the relations of this research with the circular economy, with the European trends in the digital and green transitions, as well as with the Sustainable Development Goals.

References

[1] D. Taskin, E. Dogan, M. Madaleno, 'Analyzing the relationship between energy efficiency and environmental and financial variables: A way towards sustainable development', Energy, 124045, Vol. 252, 2022.

[2] G. Di Foggia, 'Energy efficiency measures in buildings for achieving sustainable development goals', Heliyon, e00953, Vol. 4, Is. 11, 2018.

[3] M. Sovetova, J. K. Calautit, 'Thermal and energy efficiency in 3D-printed buildings: Review of geometric design, materials and printing processes', Energy and Buildings, 114731, Vol. 323, 2024.

[4] Chapter 8 - Building Automation for Energy Efficiency, Editor(s): Francesco Asdrubali, Umberto Desideri, Handbook of Energy Efficiency in Buildings, Butterworth-Heinemann, Pages 597-673, 2019.
[5] G. Qiang, S. Tang, J. Hao, L. Di Sarno, G. Wu, S. Ren, 'Building automation systems for energy and comfort management in green buildings: A critical review and future directions', Renewable and Sustainable Energy Reviews, 113301, Vol. 179, 2023.
[6] L. Vandenbogaerde, S. Verbeke, A. Audenaert, 'Optimizing building energy consumption in office buildings: A review of building automation and control systems and factors influencing energy savings', Journal of Building Engineering, 107233, Vol. 76, 2023.
[7] A. Gambarotta, R. Malabarba, M. Morini, G. Randazzo, M. Rossi, C. Saletti, A. Vieri, 'Demonstrating a smart controller in a hospital integrated energy system', Smart Energy, 100120, Vol. 12, 2023.
[8] P. Martins, A. Ramos, E. Pina, P. Váz, J. Silva, M. Abbasi, 'Smart Building Control: An Android Application for Enhanced Monitoring and Management in the Internet of Things Era', Procedia Computer Science, pp. 594-601, Vol. 238, 2024.
[9] T. Malatesta, C. Eon, J. K. Breadsell, A. Law, J. Byrne, G. M. Morrison, 'Systems of social practice and automation in an energy efficient home', Building and Environment, 109543, Vol. 224, 2022.
[10] D. R. Hax, R. K. Leitzke, A. C. S. Baptista da Silva, E. G. da Cunha, 'Influence of user behavior on energy consumption in a university building versus automation costs', Energy and Buildings, 111730, Vol. 256, 2022.
[11] C. Garcia, P. A. López-Jiménez, M.o Pérez-Sánchez, R. Sanchis, 'Methodology for assessing progress in sustainable development goals indicators in urban water systems. How far are we from the 2030 targets?', 105616, Sustainable Cities and Society, Vol. 112, 2024.
[12] S. Cairone, S. W. Hasan, K.-H. Choo, D. F. Lekkas, L. Fortunato, A. A. Zorpas, G. Korshin, T. Zarra, V. Belgiorno, V. Naddeo, 'Revolutionizing wastewater treatment toward circular economy and carbon neutrality goals: Pioneering sustainable and efficient solutions for automation and advanced process control with smart and cutting-edge technologies', Journal of Water Process Engineering, 105486, Vol. 63, 2024.
[13] R. Vadruccio, G. Salvadori, A. Tumino, 'Smart metering and Internet of Things for efficient water management', Procedia Computer Science, pp. 1498-1505, Vol. 239, 2024.

[14] A. L. Reis, M.A.R. Lopes, A. Andrade-Campos, C. H. Antunes, 'A review of operational control strategies in water supply systems for energy and cost efficiency', Renewable and Sustainable Energy Reviews, 113140, Vol. 175, 2023.

[15] S. Amertet, G. Gebresenbet, H. M. Alwan, 'Application of hyper-automation in farming – an analysis', Smart Agricultural Technology, 100516, Vol. 9, 2024.

[16] G. C. Sánchez, J. P. Monteagudo Yanes, M. Montesino Pérez, J. L. Cabrera Sánchez, A. Padrón Padrón, D. Haeseldonckx, 'Efficiency in electromechanical drive motors and energy performance indicators for implementing a management system in balanced animal feed manufacturing', 116818, Energy, Vol. 194, 2020.

[17] E. Bwambale, F. K. Abagale, G. K. Anornu, 'Towards a modelling, optimization and predictive control framework for smart irrigation', Heliyon, e38095, Vol. 10, Is. 18, 2024.

[18] C. Sheline, F. Grant, S. Gelmini, S. Pratt, A. G. Winter V., 'Designing a predictive optimal water and energy irrigation (POWEIr) controller for solar-powered drip irrigation systems in resource-constrained contexts', Applied Energy, 124107, Vol. 377, Part A, 2025.

[19] M. Schweizer, M. Stöckl, R. Tutunaru, U. Holzhammer, 'Influence of heating, air conditioning and vehicle automation on the energy and power demand of electromobility', Energy Conversion and Management: X, 100443, Vol. 20, 2023.

[20] T. Jayson, A.S.M. Bakibillah, C. Pin Tan, M.A.S. Kamal, V. Monn, J-i. Imura, 'Electric vehicle eco-driving strategy at signalized intersections based on optimal energy consumption', Journal of Environmental Management, 122245, Vol. 368, 2024.

[21] S. S. Ali Naqvi, H. Jamil, N. Iqbal, S. Khan, D.-In Lee, Y. C. Park, D.H. Kim, 'Multi-objective optimization of PI controller for BLDC motor speed control and energy saving in Electric Vehicles: A constrained swarm-based approach', pp. 402-417, Energy Reports, Vol. 12, 2024.

[22] S. Kofler, S. Jakubek, C. Hametner, 'Predictive energy management strategy with optimal stack start/stop control for fuel cell vehicles', Applied Energy, 124513, Vol. 377, Part B, 2025.

[23] C. Yin, X. Zeng, Z. Yin, 'An improved data-driven predictive optimal control approach for designing hybrid electric vehicle energy management strategies', Applied Energy, 123984, Vol. 375, 2024.

[24] E. Jungblut, T. Grube, J. Linssen, D. Stolten, 'Fuel-saving opportunities for automated vehicles: A driving cycle analysis', Transportation Research Interdisciplinary Perspectives, 100964, Vol. 22, 2023.

[25] F. Borghesan, M. Zagorowska, M. Mercangöz, 'Unmanned and Autonomous Systems: Future of Automation in Process and Energy Industries', IFAC-PapersOnLine, pp. 875-882, Vol. 55, Is. 7, 2022.

[26] A. Silvestri, D. Coraci, S. Brandi, A. Capozzoli, E. Borkowski, J. Köhler, D. Wu, M. N. Zeilinger, A. Schlueter, 'Real building implementation of a deep reinforcement learning controller to enhance energy efficiency and indoor temperature control', 123447, Applied Energy, Vol. 368, 2024.

[27] Z.-ze He, J.-yang Zhang, F.-ying Zheng, X.-jian Jin, 'Energy-efficiency-oriented optimal control for electrical environmental control system based on advanced neural network', Applied Thermal Engineering, 119635, Vol. 219, Part C, 2023.

[28] R. Abela, P. Refalo, M. Borg, E. Francalanza, 'Pneumatic Control for Sustainable Compressed Air Systems: Multi-criteria Optimisation for Energy Efficient Production', pp. 247-252, Procedia CIRP, Vol. 122, 2024.

[29] P. Righettini, R. Strada, M. Tiboni, F. Cortinovis, J. Santinelli, 'A systematic management and control methodology for high energy saving in applications equipped with hydraulic servo-axes', 105847, Control Engineering Practice, Vol. 145, 2024.

[30] V. Sousa Santos, J. J. Cabello Eras, M. J. Cabello Ulloa, 'Evaluation of the energy saving potential in electric motors applying a load-based voltage control method', Energy, 132012, Volume 303, 2024.

[31] F. Naseri, C. Barbu, T. Sarikurt, 'Optimal sizing of hybrid high-energy/high-power battery energy storage systems to improve battery cycle life and charging power in electric vehicle applications', Journal of Energy Storage, 105768, Vol. 55, Part D, 2022.

[32] D. Maamria, F. Chaplais, N. Petit, A. Sciarretta, 'Numerical optimal control as a method to evaluate the benefit of thermal management in hybrid electric vehicles', IFAC Proceedings Volumes, pp. 4807-4812, Volume 47, Issue 3, 2014.

[33] C. P. Carvajal, G. M. Andaluz, V. H. Andaluz, F. Roberti, G. 'Palacios-Navarro, R. Carelli, Multitask control of aerial manipulator robots with dynamic compensation based on numerical methods', Robotics and Autonomous Systems, 104614, Vol. 173, 2024.

[34] I. Nevliudov, M. Omarov, Y. Romashov, V. Muradova, M. Vzesnevsky, 'One approach to find optimal controls for discrete dynamic systems with numerical methods application', Advanced Mathematical Models & Applications, pp.548-564, Vol. 8, No. 3, 2023.

[35] I. Kolupaieva, I. Nevliudov, Y. Romashov, L. Tiesheva, L. Vértesy, 'Intelligent Automated Control in Accordance with Resource Efficiency Criteria toward Circular Economy Transition'. In: Kahraman, C., Cevik Onar, S., Cebi, S., Oztaysi, B., Tolga, A.C., Ucal Sari, I. (eds) Intelligent and Fuzzy Systems. INFUS 2024. Lecture Notes in Networks and Systems, vol 1089. Springer, 2024.

[36] P. J. Dias de Oliveira Evald, G. V. Hollweg, R. V. Tambara, H. A. Gründling, 'Lyapunov stability analysis of a robust model reference adaptive PI controller for systems with matched and unmatched dynamics', Journal of the Franklin Institute, pp. 6659-6689, Vol. 359, Is. 13, 2022.

[37] C. Pasquale, S. Sacone, S. Siri, A. Ferrara, 'Optimal control of electric automated buses in intercity lines: scenario-based analysis of the Pareto-optimal solutions', IFAC-PapersOnLine, pp. 37-42, Vol. 58, Is. 10, 2024.

[38] L. Lella, N. Oses-Eraso, I. Stamos, 'Pioneering a sustainable development goals monitoring framework for European regions', Ecological Indicators, 112248, Vol. 166, 2024.

[39] V.G. Boltyanski, R.V. Gamkrelidze, E.F. Mishchenko, L.S. Pontryagin, 'The maximum principle in the theory of optimal processes of control', IFAC Proceedings Volumes, pp. 464-469, Vol. 1, Is. 1, 1960.

[40] A. Ravindran, K. M. Ragsdell, G. V. Reklaitis, 'Engineering Optimization: Methods and Applications', John Wiley & Sons, Inc., Hoboken, 664 p., 2006.

[41] A. G. Mamalis, I. Nevliudov, and Y. Romashov, 'An approach for numerical simulating and processing of measured electrical signals from board sensors installed on wheeled electro-mechanical platforms', Journal of Instrumentation, article: P10006, vol. 16(10), 2021.

[42] S. Alyokhina, I. Nevliudov, Yu. Romashov, 'Safe Transportation of Nuclear Fuel Assemblies by Means of Wheeled Robotic Platforms, Nuclear and radiation safety, pp. 43-50, 3(91), 2021.

[43] P. Roux, 'Scilab: De la théorie à la pratique - I. Les fondamentaux', Éditions D-BookeR, Lille, 450 p., 2013.

[44] S. Steer, Y. Degré, 'Scilab: De la théorie à la pratique - II. Modéliser et simuler avec Xcos', Éditions D-BookeR, Lille, 557 p., 2014.

[45] G. A. Korn, T. M. Korn, 'Mathematical Handbook for Scientists and Engineers: Definitions, Theorems, and Formulas for Reference and Review', Dover Publications, Mineola, New York, 1149 p. 2000.

[46] S. Alyokhina, I. Nevliudov and Y. Romashov, 'The Feed Water Level Improved Automated Control for Steam Generators of Nuclear Power Plants', IEEE 4th International Conference on Modern Electrical and Energy System (MEES), Kremenchuk, Ukraine, pp. 1-6, 2022,

6

Climate Fitness: Decoding Urban Green Dynamics

M. Michel and J. Helnwein

University of Applied Sciences Technikum Wien, Austria
E-mail: melanie.michel@technikum-wien.at

Abstract

In tackling contemporary climate change, research and policy heavily lean on ecological data, frequently excluding the nuanced social and economic interconnections, historical inequalities, existing power structures, and competing interests that shape urban environments. This limited perception of climate adaptation and mitigation measures requires a more balanced, multidimensional assessment framework. To address this gap, the Urban Climate Reactor research team has expanded the concept of climate fitness, a term increasingly recognized in German-speaking circles but lacking a formal definition. However, this ambiguity has paved the way for a novel theoretical framework grounded in four core concepts: coping capacity, adaptability, transformability, and eco-harmony. The proposed climate fitness framework sets out three overarching goals: (1) protecting human well-being and biodiversity, (2) promoting justice, equity, and inclusion, and (3) fostering resilient urban spaces that thoughtfully integrate technological advancements. Based on an exploratory literature review and expert insights, this theoretical framework integrates a technology assessment approach to assess how effectively climate strategies align with the defined climate fitness goals. To test its applicability, a case study of photovoltaic tiles and New York's High Line Park was conducted to pinpoint the shortcomings of emerging technologies and harness the potential of existing climate projects to cultivate climate-fit cities that serve everyone. The Climate Fitness Challenges, a serious game,

DOI: 10.1201/9788743806851-6

was also developed to communicate complex research findings and engage diverse audiences in climate-fit strategies.

Keywords: Climate fitness, assessment framework, sustainable urban development, climate adaptation, climate mitigation, environmental justice, community resilience, holistic climate strategies.

6.1 Introduction

With over half of the global population expected to reside in urban areas by 2050 [1], cities are emerging as critical hubs for sustainable development and innovation as they aggregate knowledge and resources needed for climate adaptation and mitigation [2, 3]. Paradoxically, urban centers also embody many of humanity's environmental challenges [3, 4]. Increasingly frequent extreme weather events, such as storms, floods, wildfires, and heatwaves, have profound implications for urban quality of Life (QoL) [5, 6]. The latest Intergovernmental Panel on Climate Change (IPCC) Assessment Report forecasts escalating climate impacts, including disruptions to food and water security, mass displacements, heat-related health issues, infectious diseases, and a burgeoning mental health crisis due to the trauma caused by increasing visibility of and stressors caused by climatic changes [5, 7–10]. Furthermore, the impacts of climate change are not evenly distributed, further widening social disparities, with low-income neighborhoods and areas with higher levels of soil sealing facing disproportionate impacts [11–15]. Therefore, cities must urgently accelerate integrative climate adaptation and mitigation efforts [16] to safeguard both human and ecological systems.

In that context, understanding how climate change manifests itself in urban settings and decoding the roles and potential impacts of adaptation and mitigation measures are invaluable. Although there is a growing awareness that the success and reach of urban climate strategies hinge on their intersectional and multidimensional scope [17–19], current assessment models frequently have a limited ability to capture this complex, interconnected, bigger picture. For instance, evaluations of urban green spaces (UGSs) like parks tend to concentrate on ecological, aesthetic, or recreational benefits [20–23], overlooking their broader impact and potential.

To address this gap, the Urban Climate Reactor (UCR) research team proposes the concept of climate fitness as a holistic framework that goes beyond ecology to encompass social, political, economic, and technological factors. While the term climate fitness has gained recognition in the

German-speaking world [24–27], its use remains ambiguous, providing an opportunity to develop a multidimensional assessment framework that integrates four key components – coping capacity, adaptability, transformability, and eco-harmony – each of them corresponding to climate strategies that operate at varying levels of complexity, intertwining immediate actions with long-term, systemic transformations, creating a balanced approach to long-term urban resilience and "fitness."

This contribution refines and expands upon the initial concept and framework of climate fitness [24], providing a more comprehensive assessment framework. By introducing a holistic checklist of criteria, this theoretical framework enables a nuanced assessment of climate adaptation and mitigation strategies. The initial framework, while practical, compartmentalized environmental, social, health, economic, and technological dimensions [24], limiting its ability to address their interconnectedness. The revised climate fitness assessment framework overcomes these limitations, offering both guidance for new climate-related projects and evaluation of existing adaptation and mitigation projects, thus ensuring that these strategies are optimized for long-term and inclusive resilience across multiple dimensions.

The climate fitness assessment framework aims to achieve three core goals: (1) protecting human livelihoods and biodiversity, (2) advancing justice, equity, and intersectionality, and (3) fostering resilient urban spaces through effective technological integration. The interdependence of these three dimensions is profound. Biodiversity is crucial, as ecosystem resilience directly supports human well-being [28–30]. Biodiversity loss severely disrupts essential resources, such as food security, water supply, and public health [31, 32], highlighting the intrinsic connection between ecosystem services (ES) and human survival. Thus, the joint integration of these dimensions is essential to any effective climate strategy [33, 34]. Furthermore, climate strategies should rely on intersectionality through the integration of concepts and ideas related to equity and justice, as climate change disproportionately affects vulnerable communities [35–38]. Tackling social inequalities is a central building block of any long-term climate and sustainability strategy [39–42]. By strengthening well-being, cohesion, and resilience, particularly within communities most impacted by climate change, both human and ecological systems can be safeguarded at a global scale [43–46].

This theoretical framework not only fills existing gaps in climate strategy assessments but also establishes a new standard for integrated urban development. Based on an exploratory literature review and expert feedback, a technology assessment (TA) framework was crafted to evaluate how urban climate strategies align with the proposed climate fitness definition and goals. A case study focusing on photovoltaic (PV) tiles and the High Line Park (HLP) in New York City highlights the capacity of this theoretical framework to identify both limitations of emerging technologies and opportunities within existing projects, aiming to create inclusive, climate-fit cities for city dwellers, as well as fauna and flora. In tandem, the "Climate Fitness Challenge," [47] a serious game designed to enhance public understanding of complex climate strategies through transmedia communication, was developed. By simulating real-world decision-making challenges in urban planning, the game encourages diverse population groups to engage with the concept of climate fitness, bridging the gap between scientific research and public information, prompting a critical examination of how research findings are communicated, and exploring ways to enhance public awareness through gamification.

6.2 A Comprehensive Look Into the UCR Project

The UCR project, funded by Vienna's Municipal Department of Economy, Labour, and Statistics (MA 23), addresses the challenges of climate change in urban environments through comprehensive interdisciplinary and transdisciplinary assessments of both existing and emerging climate adaptation and mitigation strategies. The project aims to investigate how these measures interact to shape urban resilience and to identify potential conflicts of interest that may arise during their development and implementation. By exploring these complex interconnections, the UCR project seeks to deepen the understanding of effective urban responses to the escalating impacts of climate change.

The research team is led by Dr Susanne Schidler, a human ecologist, lecturer, and consultant specializing in gender and diversity issues. Dr Schidler brings extensive experience in managing international research projects focused on socioeconomic and ecological interactions, as well as evaluating environmental and sustainability performance. Mélanie Michel, MSc, contributes a multidisciplinary background in international relations, environmental technology, and economics. Their work examines how climate strategies can enhance urban resilience while addressing the diverse needs of various population groups, with a particular focus on creating inclusive and equitable urban environments. Michel delves into critical issues such as the intersection of gender and environment, environmental injustice and racism, and energy poverty and social justice within energy communities. Jasmin

Helnwein, MSc, offers expertise in life cycle analysis and microclimate simulations, with a background in nature conservation and biodiversity management. Her research centers on visualizing how climate change manifests in urban areas, providing insights into how cities can better utilize nature-based solutions (NBS) and technological innovations to adapt to and mitigate climate challenges, particularly in response to extreme heat.

By combining the diverse expertise of its interdisciplinary team, the UCR project aims to advance the understanding of how climate change impacts cities and to identify effective strategies to mitigate these challenges. The project focuses on identifying measures that contribute to establishing liveable and inclusive cities under the motto "climate-fit cities for all." The overarching goal is to develop a practical toolkit based on technology assessment (TA) to assist cities like Vienna in navigating multifaceted climate challenges through informed decision-making and strategic planning. This toolkit seeks to empower municipalities to implement climate-resilient strategies that optimize synergies between adaptation and mitigation measures.

6.3 Research Methodology

The methodology of this study involved several core steps, beginning with an exploratory literature review to investigate the manifestation of climate change in urban environments and the existing adaptation and mitigation strategies in place. This phase focused on assessing how different population groups are impacted by climate variability, examining both current urban adaptation initiatives and potential technologies under consideration. Special emphasis was placed on identifying strategies that are either currently implemented or planned within Vienna. Another primary objective was to contextualize and define climate fitness while examining existing frameworks where this concept has been applied.

To capture the scope of existing research, a comprehensive set of keywords was utilized. Terms such as "climate-fit" and "climate fitness," alongside their German equivalents, "Klimafit" and "Klimafitness," were employed to pinpoint research gaps. Additionally, keywords like "climate proofing" and "climate neutrality" provided insights into policy measures aimed at fostering long-term urban resilience. Broader terms, including "climate adaptation," "climate mitigation," "adaptation assessment," "mitigation assessment," and "climate strategy assessment," enabled a thorough exploration of strategies designed to prepare cities for climate impacts or address root causes of climate change.

A systematic literature review was conducted to establish a working definition of climate fitness and develop a multidimensional set of evaluation criteria. This review concentrated on the roles of UGSs, specifically parks, in addressing environmental challenges like urban heat islands (UHIs), extreme weather events, and ecological degradation while also considering the vulnerabilities of these spaces to climate impacts. Beyond their environmental contributions, the review examined the social, economic, and cultural significance of parks, analyzing their historical and epistemological roles as spaces for economic development, health promotion, education, cultural exchange, and social cohesion. Key considerations also included issues of accessibility, space allocation, social justice, gentrification, and potential conflicts of interest in the development of UGSs. Integrative approaches, such as Ferreira et al.'s socio-ecological and economic evaluation [48], Palabıyık's ecological design criteria [49], and Zhang et al.'s multi-criteria park quality assessment [50], served as foundational references for this analysis.

The findings were validated through expert consultation and transdisciplinary collaboration. The concept of climate fitness was further developed through an interdisciplinary survey at the University of Applied Sciences Technikum Wien, which gathered definitions, criteria, and indicators related to climate fitness from the Resilient Energy Systems Department, involving 41 team members. Additional insights were gathered from a presentation and discussion session at the e-nova international conference in Pinkafeld, Austria, and a climate fitness workshop held at the Joint International Congress in Adelaide, Australia, with 32 participants. During the workshop, participants collaborated to outline the roles an ideal park should fulfill, as well as criteria and indicators for assessing a park's alignment with climate fitness standards. These efforts refined the definition of climate fitness and led to the initial development of a TA framework for climate fitness.

Feedback from experts was also incorporated into the development of the "Climate Fitness Challenge," [47] a serious game showcased at the Future Green City World Congress. This board game functions as an educational and transmedia storytelling and knowledge-sharing tool designed to communicate research findings to diverse groups beyond conventional academic formats. By simulating urban planning decision-making processes, the game immerses players in the concept of climate fitness from various perspectives, including policymakers, community representatives, biodiversity specialists, technologists, and educators. This interactive approach fosters a comprehensive understanding of climate strategies by enabling players to explore options for climate-proofing cities through diverse park amenities and design features. Its successful trial at the congress underscored the game's potential for making complex research accessible and relevant through gamification.

In the final stage, the established criteria and indicators were adapted and applied to broader climate initiatives, encompassing innovative technologies like PV tiles and large-scale urban projects like New York City's High Line Park. This step culminated in a generalized assessment framework adaptable to a wide range of urban contexts and scalable to various adaptation and mitigation efforts.

6.4 Rethinking Climate Strategies Through the Lens of Climate Fitness

This section explores the concept of climate fitness by reviewing existing definitions and introducing the refined framework created by the research team. The framework comprises a detailed definition along with a set of criteria, illustrating how urban climate adaptation and mitigation strategies can extend beyond purely ecological impacts.

6.4.1 Existing climate fitness concepts

With anticipated climate shifts and projected urban population growth [1], cities are poised to take on an increasingly critical role in both climate change mitigation and the enhancement of resilience, functionality, and QoL. This growing importance highlights the need for multifaceted approaches in the development and evaluation of climate strategies. In this context, the concept of climate fitness presents valuable opportunities to address these emerging challenges.

Though relatively uncommon in English, climate fitness is a term that is gaining traction in German-speaking regions, where it appears across various sectors, including business administration, construction, and building technology. However, findings indicate that despite its frequent use, there is no standardized definition of climate fitness in German contexts [24]. Currently, the term is often applied vaguely as climate-related jargon, sometimes used interchangeably with concepts like climate neutrality, sustainability, or energy efficiency, without clear contextual explanations [24].

To capture the existing definitions and applications of climate fitness and "climate-fit" (as well as their German equivalents "Klimafitness" and "Klimafit"), a comprehensive search was conducted across three databases. This investigation exposed the fragmented understanding of these terms, as summarized in Table 6.1. Due to the high volume of results from Google Scholar, a sampling approach was taken, reviewing the first 100 search results for each term. While this approach does not cover all literature, it offers a representative view of prevailing trends and research gaps. Further refinement was done to assess the direct relevance of the term variations, with only a limited subset of results across databases directly addressing the keywords, as shown in Table 6.1.

These findings illustrate the absence of a unified definition for climate fitness, as much of the literature addresses "climate" and "fitness" separately. While some works focus on human physical and mental well-being and adaptability in response to climate change [51–56], others explore resilience and survival among various organisms, such as mammals, birds, and plants, facing environmental stressors [57–63].

Jonathan Sabolski offers a unique perspective on climate fitness, drawing on his background in marine biology and athletics to suggest adopting an athlete's mindset to tackle climate issues [64]. He advocates for incremental, achievable steps, similar to setting personal fitness goals, as a way to make progress [64]. This approach highlights the mental and behavioral changes needed to confront climate challenges, although it lacks a broader integration of interactions existing between the human and natural systems within urban spaces.

Some preliminary definitions that add clarity have emerged in climate science and spatial planning fields [26]. For example, Austria's Ministry of Agriculture defines climate-fit forests as those with well-adjusted composition and structure, enabling them to withstand current and future climate conditions [65]. Here, the emphasis is on maintaining ecosystems' natural potential. Additionally, the concept of resilience and its opposite, vulnerability, rooted in disaster sociology and geographical development studies, serves as a related framework [66]. Another example comes from the "Climate Change

Table 6.1 Results of the search query for the terms "Klimafitness," "Klimafit," "Climate Fitness," and "Climate-fit" as of October 2024 across ScienceDirect, Google Scholar, and OpenAlex.

Database	Klimafitness	Klimafit	Climate Fitness	Climate-fit	Overall, direct links to keywords
ScienceDirect	1	1	39.909	266.278	0
Google Scholar	321	315	3.180.000	1.430	63
OpenAlex	15	15	136.700	993.900	1

Adaptation by Spatial Planning in the Alpine Space (CLISP)" project, which uses a "climate change fitness" [66] criterion to assess spatial planning and land use adaptability in Alpine regions. Under this criterion, planning processes and regulations are deemed "fit" if they enhance regional resilience and adaptive capacity [25]. This interpretation of fitness includes not only ecological considerations but also socio-political factors like spatial conflicts, public engagement, political support, and climate-related economic issues [67]. Similarly, Austria's "Climate Change Adaptation Model Regions (KLAR!)" program employs climate fitness to prioritize the potential of natural ecosystems, their mitigation capabilities, and the enhancement of ES through interventions like green roofs and flood control combined with public awareness efforts [27].

In non-scientific contexts, climate fitness has been explored in creative fields, such as the exhibition "Climate Fitness: Rituals of Adaptability" [67] curated by Maite Borjabad [68, 69]. This project reimagines climate fitness artistically, examining human adaptability to climate change and exploring how physical and behavioral transformation might be necessary to confront the climate crisis and its potential impacts on society [68, 69]. This artistic exploration critiques the ambition to control both the body and the natural world, merging biological, cultural, and social aspects and further expanding the concept beyond traditional scientific approaches [68, 69].

Existing definitions of climate fitness generally emphasize ecological and energy-related aspects, overlooking socioeconomic, political, health, and technological dimensions that are essential for developing resilient, future-ready urban environments. The lack of a standardized, multidimensional definition reveals a gap in the literature. On one level, climate fitness might appear as another emerging "green" buzzword, similar to terms like sustainability or climate neutrality. However, this lack of specificity presents an opportunity to establish a well-rounded conceptual framework that evaluates the socio-ecological compatibility of climate initiatives. This study seeks to elevate the notion of climate fitness from a fragmented usage to a multidimensional concept, aiming to guide climate efforts in urban settings and embed the term as a comprehensive tool in environmental discussions.

6.4.2 A new perspective on climate fitness

Research findings suggest that the current understanding of climate fitness is often simplistic and lacks depth, failing to capture the complex interactions and dependencies. For example, studies on urban parks typically examine only a few dimensions limiting the system boundaries to being socio-technical, socio-ecological, or eco-economic. Specific studies illustrate this focus: Dong et al. [21] explore park accessibility indices, Campbell et al. [20] investigate the social significance of parks, Silva et al. [22] assess the economic value of historic parks, and Suhardono et al. [23] examine ecological functions within park environments.

Moreover, Lesnikowski et al. [70] stress the need for a comprehensive approach to climate policy, Van Den Bergh et al. [19] point to the lack of integration in climate strategies, which limits the potential for positive synergies, while Rylenius and Hamza [17] argue for cohesive climate adaptation methods to avoid maladaptive consequences. Sethi et al. [18] further highlight the fragmented nature of knowledge and planning systems for climate action, advocating for stronger collaboration between researchers, policymakers, and urban institutions to improve adaptation and mitigation efforts. Collectively, these studies underscore the importance of understanding the roles, impacts, and opportunities within emerging climate strategies, as well as identifying limitations in current projects that could be addressed.

Given these insights, it is essential to broaden the concept of climate fitness due to the lack of holistic assessment frameworks that meet the complexities of urban areas facing climate challenges. The foundational step in crafting a broader, intersectional definition of climate fitness involves moving beyond strictly ecological perspectives to include various facets of human society. Climate fitness, in this context, can be defined as a system or community's ability to withstand and adapt to shifting climate conditions while simultaneously delivering benefits to both the environment and society [24]. Key elements of climate fitness are as follows:

- Coping capacity refers to a system's resilience and robustness, focusing on its ability to reduce the adverse effects of climate change. This concept encompasses preventive and proactive climate measures that bolster urban areas against external climate impacts by promoting low-stakes, easily integrated strategies. Examples include stormwater management techniques such as permeable pavement [71], rain gardens [72], and versatile green-blue-grey infrastructures [73, 74]. These relatively cost-effective solutions can be incorporated into existing urban structures, delivering quasi-immediate benefits.
- Adaptability builds upon coping capacity by emphasizing a system's ability to adjust and thrive amid evolving climate conditions. This component highlights the importance of flexible, responsive solutions that

evolve based on feedback from stakeholders like local communities and municipalities, as well as environmental changes. Medium-stakes measures in this category might include climate-responsive building codes that integrate traditional architecture [75], passive energy strategies [76], natural ventilation systems [77], and flood-resistant designs [74], as well as other climate-responsive strategies [78]. These approaches contribute to safeguarding building infrastructure against climate risks, thereby supporting its long-term durability.

- Transformability addresses the need for transformative shifts when existing technical, socio-political, ecological, or economic systems become unsustainable or ineffective. This high-stakes approach involves proactively rethinking and redesigning systems in response to major climatic shifts or emerging research insights. Transformability requires substantial adjustments that ensure communities and their members are well-prepared for future climate challenges and opportunities. An example of this level of change would be the complete transition to sustainable energy systems and low-carbon economies. Such a shift necessitates extensive policy reforms, infrastructure redesigns, and the development of new skills and industries to support social and environmental well-being [79, 80].
- Eco-harmony focuses on the integration of human activities with ecological processes, ensuring that human actions do not disrupt but instead complement and enhance the natural world and its ES. This dimension emphasizes a symbiotic relationship between human society and the environment, aiming for a balance that promotes the health of both without compromise. Medium-stakes, proactive measures under eco-harmony include the development of urban green corridors that connect various UGSs. These corridors can mitigate extreme urban heat [81], improve air quality [82], manage stormwater [83], and boost the well-being and QoL of urban residents [84]. Additionally, these green networks support biodiversity and foster cultural exchange within urban communities [85–87].

The initial iteration of the climate fitness assessment framework applied a segmented approach, evaluating the ecological, social, health, economic, and technological aspects independently [24]. Although these dimensions were addressed individually, they were analyzed in isolation, as illustrated in Figure 6.1. This limits the framework's effectiveness in capturing the complex interactions inherent in urban climate strategies.

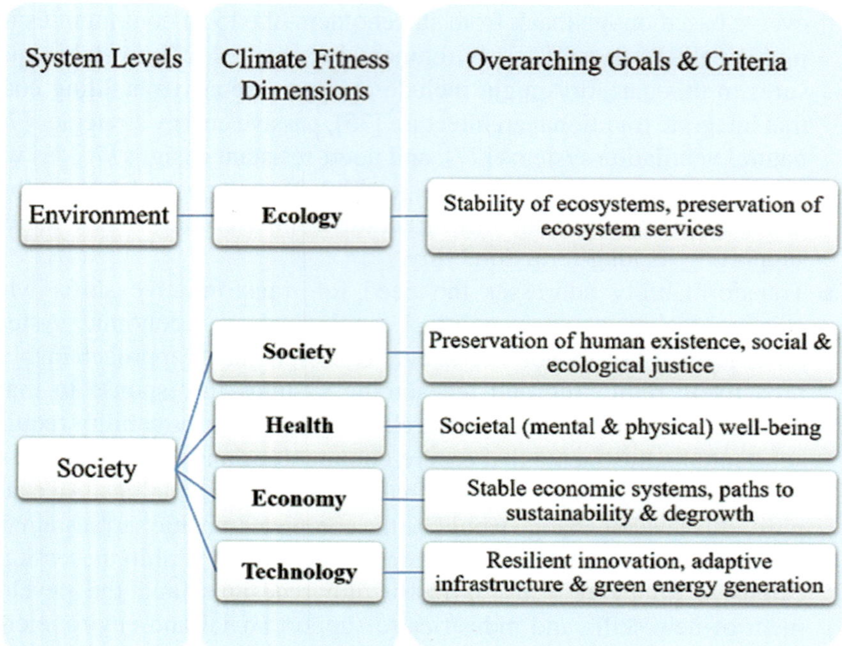

Figure 6.1 Initial version of the climate fitness dimensions, along with the defined overarching goals and criteria that form the assessment framework.

In contrast, the revised framework prioritizes the interconnectedness of these dimensions. Three overarching goals collectively shape a comprehensive framework for addressing climate challenges, balancing ecological integrity, public health, social equity, and technological integration. These goals include: (1) protecting human well-being and biodiversity, (2) promoting justice, equity, and intersectionality, and (3) fostering resilient urban transformation alongside holistic approaches to technology use. These overarching goals, along with their core criteria, are illustrated in Figure 6.2.

The primary objective of (1) safeguarding human livelihoods and biodiversity is to promote the well-being of both natural and human communities, rooted in the principle of eco-harmony. This goal prioritizes the protection and enhancement of the livelihood of humans, plants, and animals, aiming for preservation that exceeds basic survival needs. Key components of this goal include (1.1) improvements to urban QoL, life satisfaction, and the local microclimate, with a focus on how climate initiatives reduce the adverse effects of urbanization and climate change. This includes addressing issues

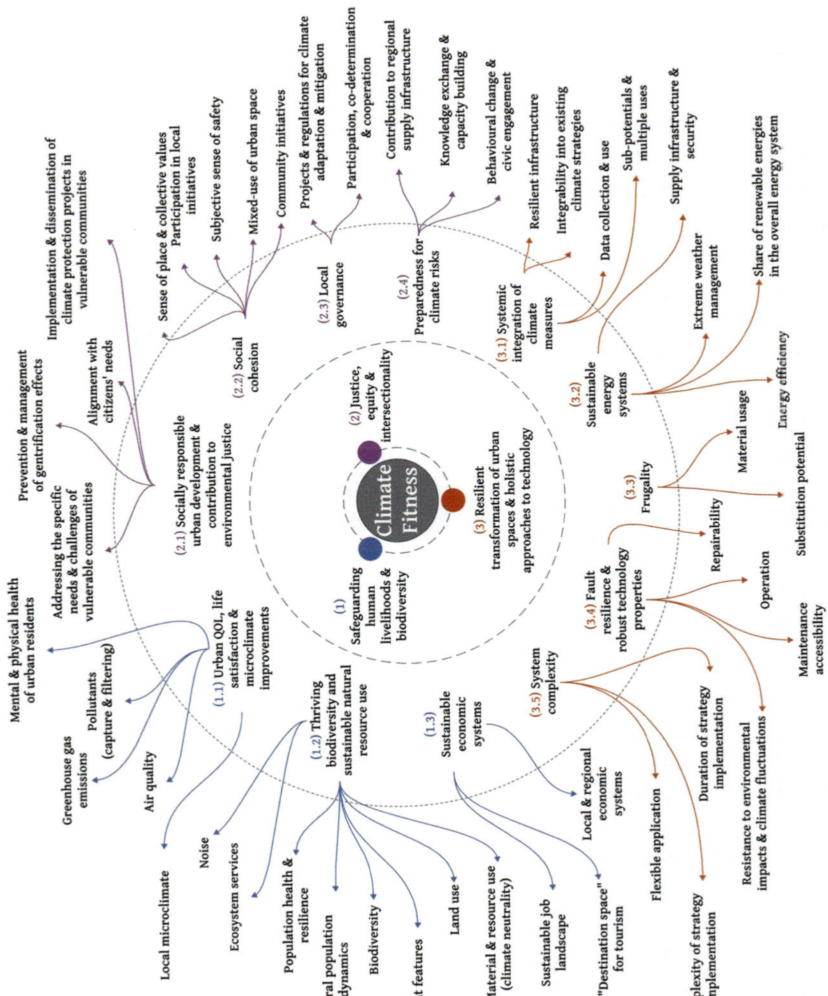

Figure 6.2 Essential climate fitness criteria associated with each defined overarching goal.

such as noise reduction, air quality improvement, greenhouse gas (GHG) emissions reduction, and supporting both mental and physical health. Additionally, the framework considers the local microclimate by assessing the cooling impacts of implemented measures. Another focus is (1.2) promoting biodiversity and sustainable use of resources, which involves evaluating the circularity of practices in terms of land, materials, and other resources and understanding their impacts on ecological balance, species diversity, and ES. Finally, (1.3) sustainable economic systems are evaluated, emphasizing efforts to bolster local and regional economies by creating sustainable job opportunities and enhancing the area's appeal for tourism, positioning it as an attractive destination.

The second major goal, with a focus on (2) justice, equity, and intersectionality, offers a unique perspective by emphasizing a form of Social Return on Investment (SROI) that values non-financial outcomes. This approach focuses on enhancing social well-being and equity by balancing "social, environmental, and economic costs and benefits" [90, 91]. This balance aims to predict the societal impact of climate measures, assessing whether they contribute to the overarching objective of climate fitness. This socially focused goal is structured around four main criteria to evaluate climate strategies for equity and resilience. The first criterion, (2.1) socially responsible urban development and environmental justice, assesses how effectively climate initiatives address the needs of diverse population groups, particularly vulnerable communities impacted by historical and structural inequities. This includes efforts to counter environmental injustice through reparative measures and strategies to mitigate gentrification. The second criterion, (2.2) social cohesion, examines the benefits and accessibility of climate measures for the general public, as well as their impact on perceived safety, community connectedness, and levels of civic engagement. The third criterion, (2.3) local governance, focuses on fostering community-centered, inclusive, and transparent decision-making processes. It also emphasizes collaboration between governmental agencies, non-governmental organizations (NGOs), and localresidents to amplify citizen representation and ensure that climate measures align with the needs and aspirations of both current and future generations. The final criterion, (2.4) climate risk preparedness, considers how well climate measures support knowledge sharing, capacity building, public engagement, and resilient social safety networks, which collectively strengthen society's ability to adapt to the socio-ecological changes driven by climate change.

A pivotal inclusion in the climate fitness assessment framework is the integration of technological aspects, recognizing that adaptation and mitigation efforts often possess unrealized potential beyond their conventional applications. Urban parks, for instance, can contribute to sustainable energy generation and act as platforms for demonstrating renewable technologies, fostering public awareness, educational opportunities, and acceptance of sustainable solutions [88–91]. Effective city planning for climate resilience necessitates a combined approach that incorporates green and blue infrastructure with technological elements, respectively grey infrastructure, which includes advancements and innovative solutions essential for building future-ready cities. Balancing ecological and technological components is thus critical to ensuring that urban environments become as adaptive and resilient as natural ecosystems.

The third overarching goal, therefore, emphasizes (3) resilient urban transformation and integrated technology usage, highlighting how technology can play a pivotal role in addressing climate-related challenges by focusing on adaptability, efficiency, and prudent resource use. This goal comprises five specific criteria supporting the resilient transformation of urban areas. First, (3.1) systemic integration assesses the capacity of climate measures to contribute to robust urban infrastructure, emphasizing the importance of merging these measures seamlessly with existing systems, technologies, and strategies to endure climatic changes while maximizing multifunctional and ancillary benefits. Effective data monitoring and application further enhance management efficiency, urban QoL, and informed decision-making processes. The criterion of (3.2) sustainable energy systems evaluates the infrastructure's reliability and its potential to enhance urban energy efficiency. In (3.3) frugality, the focus shifts to optimizing resource use, aiming to reduce material consumption without compromising functionality, thereby avoiding or substituting resource-heavy solutions. Criterion (3.4) addresses fault resilience and the robustness of technological solutions, evaluating the climate measure's durability under environmental pressures and considering ease of repair and maintenance. Finally, (3.5) system complexity is assessed, with an emphasis on adaptability to meet the unique characteristics of various urban areas, as well as on the skill requirements, complexity, and time needed for implementation – all factors that influence effectiveness and public support.

In summary, this climate fitness framework serves as a foundational tool for fostering adaptable and resilient urban communities. It facilitates the evaluation of current climate strategies, highlights areas requiring improvement to reach climate fitness goals, and supports the development of environmental initiatives that promote long-term resilience.

6.5 Applying the Climate Fitness Framework: Case Studies in Urban Green Infrastructure and Emerging Technologies

A preliminary application of the climate fitness assessment framework was conducted on two contrasting cases: an established urban greening initiative, New York City's HLP, and an emerging technology, PV floor tiles.

The HLP is a notable example of urban transformation [92], repurposing obsolete railway infrastructure into an elevated linear park that revitalizes city space and brings adaptive reuse to the urban environment [93–95]. In contrast, PV floor tiles represent an innovative technology with diverse potential applications, such as illuminating pathways, bike lanes, and parking areas, although its maturity and long-term viability are still under exploration [96, 97]. Consequently, the main challenge encountered was data availability. Analyzing a well-documented project like the HLP, which offers substantial data and concrete results, was starkly different from assessing a less-established, evolving technology with limited empirical data.

The lack of available data for PV floor tiles made it challenging to evaluate specific indicators in the framework. For instance, there is currently no information on waste disposal management for these tiles, as the technology has only been in use since 2016 [97], and thus data on their durability, maintenance needs, and disposal processes are limited. Additionally, PV floor tiles have not yet demonstrated clear economic benefits, which limits their broader application and raises questions about their potential [97]. Due to these data gaps, assumptions based on conventional rooftop PV systems were necessary to evaluate their potential impact on gentrification, community empowerment, and regional value creation alongside broader economic effects, as well as long-term effects on resource use and sustainability [98–101]. This exercise highlighted the framework's capacity to pinpoint critical information gaps, underscoring the need for supplementary studies, such as community engagement frameworks and feasibility assessments, to determine optimal integration methods for PV tiles within urban infrastructures.

Conversely, applying the climate fitness framework to the HLP exposed notable inconsistencies between the project's intended goals and its actual outcomes, particularly concerning social equity. Although the HLP is widely regarded as a sustainable urban park [102, 103] that promotes the idea of resilience for nature, the city, and its inhabitants [104], it has inadvertently contributed to gentrification in the surrounding area. This phenomenon has displaced long-term, low-income residents [105, 106], thus eroding the social fabric [103, 107] rather than enhancing social support structures, which conflicts with the framework's criteria for socially responsible urban development and environmental justice. To address such issues, the framework suggests implementing anti-gentrification measures, such as rent stabilization and affordable housing initiatives, in collaboration with governmental agencies. It also advocates for a "just green enough" approach [108, 109], which promotes smaller, widely distributed green projects in under-resourced neighborhoods. This strategy helps ensure that ecological benefits are accessible to all residents, particularly those most affected by climate change, as well as those historically and structurally excluded from the advantages of climate strategies, thereby aligning with the core principles of climate fitness.

6.6 Knowledge Dissemination Through Serious Game: The Climate Fitness Challenge

"Evidence does not speak for itself – it must be communicated" [110]. Yet, traditional methods for communicating research findings, such as scientific papers, reports, or conference proceedings, can create barriers to accessibility, as their format and language primarily cater to audiences that are well familiarized with academic frameworks [111]. This incidental exclusion hinders research findings from reaching a broader, interdisciplinary audience, ultimately limiting the real-world impact and public engagement with the research.

In that context, gamification emerges as a powerful strategy to address these challenges by transforming complex information and knowledge into interactive, engaging, and user-friendly formats to optimize learning and interaction [112, 113] by incorporating game elements, such as roleplays or challenges, in contexts that are not related to gaming [114, 115]. From there, one can assume that gamification opens up avenues for immersive learning environments that allow diverse audiences, who might otherwise be excluded from traditional academic and scientific discourse, to engage with research outcomes.

Complementing gamification, transmedia provides an additional avenue to widen the reach and impact of research dissemination by utilizing differing communication channels and media formats [116], each of them serving as a targeted touchpoint for exchanging knowledge [117]. Indeed, transmedia approaches have the potential to reach varied population groups by tailoring the content to diverse preferences, enabling research engagement across different social, cultural, educational, and professional contexts [118].

The Climate Fitness Challenge, a serious game grounded in real-life urban challenges currently emerging in Vienna, extends beyond traditional methods of communication for project outcomes and research results. This board game, designed around role-play and structured debate, tackles issues such as equitable distribution and access to UGSs, strategies for urban heat mitigation and improved QoL, community cohesion, and ensuring safety in public spaces. By immersing participants in scenarios requiring collaborative problem-solving and decision-making, the game facilitates hands-on engagement with the concept of climate fitness and turns participants into active co-creators of ideas. Designed for versatility, the Climate Fitness Challenge can be implemented across various engagement contexts: in educational institutions to introduce students to urban planning and environmental policy, at public information sessions to involve residents in local planning discussions, and in community workshops to collect insights into public needs and concerns.

By merging gamification and transmedia approaches, research dissemination has the opportunity to become more inclusive and impactful, encouraging participatory dialogue and knowledge-sharing outside the academic realm.

6.6.1 Object of analysis: Westbahnpark

The Westbahnpark, a park-related project proposal for Vienna's 15th district, known as Rudolfsheim-Fünfhaus, has garnered attention amid rising concerns over the UHI effect and has become a focal point of a public petition [119–123]. This yet-to-be-realized project served as the conceptual foundation for the Climate Fitness Challenge due to its vision of future-oriented planning for liveable cities, wherein parks are designed as urban cooling agents and deployed as proactive climate adaptation and mitigation strategies, as imagined by the WESTBAHNPARK.JETZT initiative.

The WESTBAHNPARK.JETZT initiative is a movement launched in 2020 by Hannes Gröblacher, Lilli Lička, and Karoline Seywald – core members of the "Büro für lustige Angelegenheiten (BLA, Office for fun affairs)" [124], an organization made up of landscape architects, professors, and designers, among others, focusing on developing new approaches to conflicts of interest manifesting themselves within urban public spaces.

The BLA's manifesto pledges for the establishment of "meaningful open spaces in an open society" [124]. Furthermore, the initiative advocates for the establishment of high-quality, easily accessible, and inclusive recreational spaces that are designed with the community's interests in mind, using their input and approval as guidance. The emphasis lies in the belief that the mere provision of facilities and amenities is inadequate in creating a vibrant public space, as it takes more than benches and picnic tables to create vibrant public spaces that challenge and solve existing conflicts of interest. As put by the BLA, "Equipment is not the solution" [124]. Their demands addressed to city officials revolve around effectively connecting the urban fabric and improving urban QoL by enhancing the role of public spaces. This is to be achieved through the removal of existing barriers that compromise the availability and accessibility of vital public space, as well as by preserving the "natural" state of these spaces through the avoidance of excessive bureaucratic practices that promote rapid urbanization and soil-sealing, leading to the loss of ecological and historical heritage.

Consequently, the BLA's vision entails the creation of cities that not only fulfill basic infrastructural requirements but also foster socially compatible, diverse, thriving, and climate-friendly public spaces. Central to their principles is the establishment of parks that are vast in size, even within dense areas, such as the proposed Westbahnpark – a project proposal of a linear "High-End-Wild-Line" [125] park of around 6.4 ha spanning a 1.2 km along the tracks of Vienna's Westbahnhof (Vienna West Station) [125], one of the city's most prominent train stations.

Adhering to the motto of "a park for everyone" [125], the primary aim of this initiative is to preserve tolerable urban climatic conditions for present and future generations through socially oriented and environmental design principles. Consequently, of the total surface area, 70,000 m^2 is designated to remain unsealed [125], serving as a fresh air corridor and stormwater management system while also offering visual enhancements to the city. Furthermore, the park initiative not only aims to bring green spaces closer to the residents of the 15th district but also dismantle the barrier created by railway tracks by effectively establishing connections between the 6th, 7th, and 15th districts through walkways and cycling routes. Not only does this strengthen the social fabric of the community, but it also promotes micromobility – a crucial step toward creating sustainable cities, of which walkability is the basis [126].

Given the initiative's emphasis on integrating urban resilience and social-ecological justice, the Westbahnpark initiative is representative of a "climate-fit" project. As noted by meteorologist and climate scientist Helga Kromp-Kolb, making Vienna climate-fit necessitates holistic planning strategies and projects, such as the Westbahnpark that protect and create fresh air corridors, recreational areas, and biodiversity support areas [125].

Important to note is that the initiative emerged in response to the unique socioeconomic, demographic, and environmental dynamics of the area. Vienna's 15th district is recognized for its diversity, high concentration of buildings and residents, scarcity of green recreational spaces, and economic vulnerability. With 75,635 inhabitants living across 3.9 km^2 [127] Rudolfsheim-Fünfhaus stands out as Vienna's third youngest, with an average age of 39.7 [127], exhibiting the highest foreigner ratio of Vienna with 43.1% [127], as well as the lowest average gross income, recorded at 20.042 EUR per year [127] combined with an unemployment rate of 16.1% which is also located at the upper echelon, with 85% of Vienna's districts having higher employment rates [127].

The district's recorded population density varies between 19,302 inhabitants per km^2 [127] and 20,474 inhabitants per km^2 [125]. In relation to other districts, the land use distribution of the 15th district distinguishes itself by its stark urbanization. Especially in comparison to other peripheral districts, Rudolfsheim-Fünfhaus is one of the more densely populated districts, with built land accounting for 56% of the total surface area, traffic infrastructure occupying 35%, and green spaces comprising only 9%, leading to each inhabitant being allocated just over 4 m^2 of green leisure space, which is far-off from the city's goal of providing 8 m^2 to each inhabitant [125].

Based on findings from the environmental consultancy company ECOTEN, the 15th district has an average Urban Heat Vulnerability Index (UHVI) of 0.7 [128], which represents a high to very high vulnerability. This makes the district exceedingly sensitive to, respectively, less resilient to extreme heat, limiting the area's ability to adjust to and recover from these climatic impacts. Thus, it is safe to conclude that the 15th district constitutes one of Vienna's climate change hotspots.

The case of the 15th district serves as an illustrative example of environmental injustice, namely the unequal access to high-quality green leisure spaces among different social groups and districts within Vienna. The distribution of UGSs, such as parks, in cities pertains not only to urban design and planning but also to the intrinsic interplay between socioeconomic and political factors that shape these disparities. Indeed, urban landscapes and their UGSs are inherently political, reflecting the current zeitgeist, thus visualizing broader tensions and conflicts over power, access, regulations, and representation [129, 130]. Consequently, social hierarchies, political agendas, and cultural codes [130, 131] distinctively shape the urban landscapes and, by extension, determine the characteristics and attributes of UGSs [132], thus leading to the reoccurrence of historically rooted power relations and social disparities in access to UGSs for vulnerable groups.

Given the distinctive socio-demographic and socioeconomic fabric of Vienna's 15th district, the proposed Westbahnpark project presented an ideal foundation for developing a serious game that addresses contemporary challenges within a realistic and localized context. The UCR team's prior work on the climate fitness assessment framework, which focused solely on UGSs, provided a valuable basis for the game development.

6.6.2 Game design: Learning objectives, structure, and mechanics

The Climate Fitness Challenge, as depicted in Figure 6.3, is designed to immerse players in the multifaceted decision-making processes involved in designing resilient urban spaces. This serious game aims to educate players on the importance of parks in urban spaces, emphasizing specific design considerations essential to reaching climate fitness and demonstrating the complexities of balancing diverse needs and interests, such as community priorities, budget constraints, governmental policies, and ecological stability. Through the gameplay experience, players are not only introduced to the multiple roles that parks play in urban areas but also to the notions of climate fitness, coping capacity, adaptability, transformability, and eco-harmony, highlighting how these values can inform the design of climate adaptation and mitigation projects.

The primary learning objectives are to support the development of a comprehensive understanding of urban climate resilience, stakeholder dynamics, and strategic planning as the game introduces players to the impacts of various design choices. By assuming the roles of environmental specialists, community liaisons, government officials, technology experts, and educators, players gain insights into the competing priorities that influence urban planning. The game also encourages critical thinking and strategic resource allocation as players navigate budgetary constraints, evaluate trade-offs, and conceptualize climate fitness through the perspective of their assigned roles. Through their unique character goals, players prioritize actions that advance their objectives while engaging in teamwork and negotiation, fostering collaborative problem-solving and forward-thinking planning. The Climate Fitness Challenge, thus, fosters a holistic understanding of urban climate strategies, guiding players in developing an intuition for recognizing essential design elements and synergies that constitute a climate-fit urban park.

Figure 6.3 Content of the climate fitness challenge (Source: authors).

The Climate Fitness Challenge was purposefully designed to ensure an immersive, hands-on experience that fosters in-depth discussions. Thus, the game is structured as a board game featuring 15 cards that guide the storyline, of which 13 cards present distinct challenges related to designing a climate-fit UGS. For each scenario, three response options are provided, each with an associated cost varying between 5 and 15 currency units, along with considerations outlining the advantages and potential drawbacks of each option to guide the discussion:

- The "record heatwaves" scenario prompts players to address the impacts of UHIs, mainly focusing on strategies to keep park spaces cool. Options for response include implementing high-albedo materials to reflect sunlight, installing green-roofed seating areas for artificial shading, or planting large trees to provide natural shade.
- In "connectivity and mobility," players confront the challenge of enhancing access to and within the park. Decision points include making the park a pedestrian-only area, creating mixed pathways for micromobility devices, or establishing a park-and-ride facility.
- The "park safety" scenario addresses the need to create a secure and inviting environment, especially for vulnerable groups affected by verbal

and physical violence in public spaces. Players must choose from options such as restricted hours with perimeter fencing, unrestricted access with enhanced lighting and emergency call boxes, or extended hours with partial fencing and closed-circuit television (CCTV) cameras.

- The "design features" scenario allows players to select amenities that enhance the park's attractiveness and functionality, such as a dog park, an age-inclusive playground, or a natural swimming pond.
- In the "nature and community" scenario, players focus on promoting biodiversity as a means to foster community engagement. Options include establishing community garden spaces, incorporating native plants to support local wildlife, or using decorative plants to create a visually appealing landscape.
- "Education and community engagement" challenges players to build community awareness and foster environmental stewardship. Options include a "godparent" program where residents care for specific plants, educational signage to inform visitors about local biodiversity, or school partnerships to promote environmental education.
- The "comfort and well-being" scenario directs players to consider restroom facilities, balancing sustainability, safety, and comfort. Options range from installing simple public urinals to compost toilets using woodchips and wheelchair-accessible smart toilets with automated lighting.
- In the "green energy" scenario, players are tasked with integrating renewable energy solutions into the park, aiming to both visualize sustainable energy generation and use, as well as to normalize its acceptance among the community. Choices include solar-powered park lighting, shaded seating areas with integrated PV panels, and a walkway with integrated PV tiles that light up at night.
- "Climate preparedness" presents the need to future-proof the park against climate-related challenges. Options include rain gardens to manage stormwater, combined water and misting stations to enhance visitor well-being and health, or creating a dense pocket forest to enhance biodiversity and shade.
- "Cultural exchange" focuses on utilizing the park as a space to foster community engagement and celebrate cultural diversity. The given options are organizing a weekly local produce market, hosting quarterly cultural fairs, or developing a heritage trail that celebrates the district's diversity.

The remaining two cards introduce unexpected changes, requiring the players to adapt their strategy and adding a layer of complexity to the game, requiring players to realign their strategy and priorities to meet the new constraints:

- The "budget cut" scenario introduces the reality of fluctuating financial resources, illustrating how election results and changes in government can shift political priorities and affect resource availability, highlighting the importance of adaptive resilience in project planning.
- The "policy shift" scenario introduces political complexities by requiring players to adjust their plans in response to a new policy that reduces the park's available space. This prompts players to discuss what planned projects can be canceled out without jeopardizing the overall climate fitness of this UGS, reflecting the impact of evolving policy agendas on urban planning.

Regarding the mechanics, each player is randomly assigned a character role, embodying a distinct perspective and encouraging players to advocate for their character's values, thus bringing unique insights to the decision-making process. The game begins with an initial budget of 200 units of a fictional currency, which they must strategically allocate to design a park that fulfills the principles of climate fitness. This dynamic interplay of responsibilities, resource management, and collaborative decision-making allows participants to gain a nuanced understanding of the complexities of the concept of climate fitness.

6.6.3 Educational value: Preliminary evaluation

The Climate Fitness Challenge was trialed at the Future Green City World Congress 2024 in Utrecht, Netherlands, with 10 participants divided into two game groups, as depicted in Figure 6.4. Here, it demonstrated significant potential as an educational and transmedia communication tool to make complex research results accessible to a broader audience through gamification.

Feedback from conference attendees highlighted the game's strength in generating critical discussion around sustainable urban development and climate fitness, as well as trade-offs and the complexity of balancing competing values and priorities, alongside budget limitations. Participants appreciated the structured yet open-ended nature of the game, which allowed for meaningful debate and in-depth discussion on park design and possible synergies,

(a) (b)

Figure 6.4 Photographs of the Climate Fitness Challenge being trialed at the Future Green City World Congress 2024 in Utrecht, Netherlands.

demonstrating the game's potential as a valuable educational tool. The interactive role-playing aspect was also well-received, as participants appreciated the opportunity to embody diverse perspectives. This approach fostered a deeper understanding of the nuanced decision-making that is required in urban planning.

6.7 Discussion, Conclusion, and Outlook

Our analysis, supported by related studies [17–19, 70, highlights the urgent need for an integrated, multidimensional approach to urban climate strategies, challenging the traditional ecological focus of siloed approaches. These conventional models often fail to account for intersectionality and the potential interrelations or conflicts of interest that could foster positive or negative synergies. The climate fitness assessment framework bridges these gaps by offering a clear definition that emphasizes a system's coping capacity, adaptability, transformability, and eco-harmony, coupled with a comprehensive set of criteria organized into three overarching goals.

The primary goal, safeguarding human livelihoods and biodiversity, prioritizes enhancing urban QoL by creating environments where people and biodiversity can flourish. This includes establishing sustainable, circular economic systems that operate with resource efficiency and promote both species diversity and ecological health. The second goal, promoting justice, equality, and intersectionality, focuses on reinforcing the social fabric and resilience of urban communities through fair governance that prioritizes representation, inclusive design, and reparative measures. The final goal, the resilient transformation of urban spaces and integrated technology use, highlights the critical role of technological innovations in supporting urban adaptation and mitigation efforts, enhancing resilience to environmental stresses, and improving energy efficiency.

The climate fitness framework is designed as a robust tool for policymakers, urban planners, and project managers to systematically assess and improve urban climate efforts. By identifying their strengths and limitations, the framework directs attention to critical areas for intervention, guiding cities toward balanced ecological sustainability, social equity, economic resilience, urban durability, and technological advancement.

The results from this preliminary application of the climate fitness framework suggest that it holds promise as a tool for recognizing areas for enhancement in urban development projects. For emerging technologies like PV tiles, the framework stresses the need for rigorous data collection and feasibility analysis before wide-scale deployment, ensuring that climate strategies are founded on principles of long-term viability [96, 97]. For established initiatives like the HLP, the framework reveals the complexities of replicating large-scale projects across diverse urban settings, advocating for adaptable solutions that support broader implementation [92, 133, 134]. While the social impacts of the HLP are well-documented in existing literature [94, 106, 132, 135–140], our application of the framework provides actionable insights for bolstering resilience through rent stabilization, affordable housing programs, and increased community engagement to mitigate displacement risks. By shifting the focus from tourism-centered development to equitable benefit-sharing, the framework underscores the importance of community-centered design and accessibility in UGS planning. These case studies illustrate how the climate fitness assessment framework can identify discrepancies and reveal areas requiring additional measures within current and emerging climate initiatives. Building on these findings, future research will apply the climate fitness framework across a range of new technologies and established climate measures to verify its adaptability.

Lastly, the Climate Fitness Challenge represents an attempt by the UCR research team to disrupt traditional methods of knowledge-sharing and research outcome communication and to counteract "hierarchical processes of knowledge transfer" [111]. Studies suggest that gamification and transmedia storytelling can notably enhance engagement with diverse audiences by providing multiple engagement points for interaction through a plethora of media formats and game elements [113, 115, 117]. In this context, the Climate Fitness Challenge functions as an accessible, interactive, and hands-on tool for exploring complex urban challenges, actively involving individuals outside of the academic sphere in co-creating innovative ideas for holistic urban resilience. By accompanying traditional formats such as academic papers and conference proceedings with this serious game, the UCR research team seeks to diversify its dissemination strategy and broaden its overall reach. However, to fully assess the educational value of the Climate Fitness Challenge, broader application across various settings, such as schools, universities, community workshops, and public engagement events, will be essential.

References

[1] P. Gerland, S. Hertog, M. Wheldon, V. Kantorova, D. Gu, G. Gonnella, et al., 'World Population Prospects 2022: Summary of Results', United Nations, 2022.

[2] T. Kuramochi, S. Lui, N. Höhne, S. Smit, M. J. de Villafranca Casas, F. Hans, et al., 'Global Climate Action from Cities, Regions and Businesses: Impact of Individual Actors and Cooperative Initiatives on Global and National Emissions', NewClimate Institute, Data-Driven Lab, PBL, German Development Institute (DIE), Blavatnik School of Government, University of Oxford, 2019.

[3] K. R. Slater, J. B. Robinson, 'Transformational Climate Actions by Cities', Buildings and Cities, vol. 4, pp. 74–82, 2023. [Online]. Available: https://doi.org/10.5334/bc.285.

[4] M. Pregnolato, 'Infrastructure Opportunities for the Resilience of Tomorrow's Cities', Copernicus Meetings, 2021. [Online]. Available: https://doi.org/10.5194/egusphere-egu21-2215.

[5] K. Calvin, D. Dasgupta, G. Krinner, A. Mukherji, P. W. Thorne, C. Trisos, et al., 'Climate Change 2023: Synthesis Report. Contribution of Working Groups I, II and III to the Sixth Assessment Report of the Intergovernmental Panel on Climate Change [Core Writing Team,

H. Lee and J. Romero (eds.)]', IPCC, Geneva, Switzerland, 2023. [Online]. Available: https://doi.org/10.59327/IPCC/AR6-9789291691647.

[6] K. L. Ebi, J. Vanos, J. W. Baldwin, J. E. Bell, D. M. Hondula, N. A. Errett, et al., 'Extreme Weather and Climate Change: Population Health and Health System Implications', Annual Review of Public Health, vol. 42, pp. 293–315, 2021. [Online]. Available: https://doi.org/10.1146/annurev-publhealth-012420-105026.

[7] M. O'Donnell, L. Palinkas, 'Taking a Trauma and Adversity Perspective to Climate Change Mental Health', European Journal of Psychotraumatology, vol. 15, p. 2343509, 2024. [Online]. Available: https://doi.org/10.1080/20008066.2024.2343509.

[8] M. Mayrhofer, M. Ammer, 'Climate Mobility to Europe: The Case of Disaster Displacement in Austrian Asylum Procedures', Frontiers in Climate, vol. 4, 2022. [Online]. Available: https://doi.org/10.3389/fclim.2022.990558.

[9] R. C. Ghosh, C. Orchiston, 'A Systematic Review of Climate Migration Research: Gaps in Existing Literature', SN Social Sciences, vol. 2, p. 47, 2022. [Online]. Available: https://doi.org/10.1007/s43545-022-00341-8.

[10] W. Leal Filho, L. Ternova, M. M. Fayyaz, I. R. Abubakar, M. Kovaleva, F. K. Donkor, et al., 'An Analysis of Climate Change and Health Hazards: Results from an International Study', International Journal of Climate Change Strategies and Management, vol. 14, pp. 375–98, 2022. [Online]. Available: https://doi.org/10.1108/IJCCSM-08-2021-0090.

[11] D. Regas, B. Shane, T. Schottland, L. Hwang, W. Klein, R. Chapman, et al., 'The Power of Parks to Address Climate Change - A Special Report', Trust for Public Land, San Francisco, CA, 2022.

[12] M. Lalonde, 'Opinion: Climate Change is Here, but It's Not the Great Equalizer We Thought It Might Be', Pique Newsmagazine, 2021. [Online]. Available: https://www.piquenewsmagazine.com/piquen-yer-interest/opinion-climate-change-is-here-but-its-not-the-great-equalizer-we-thought-it-might-be-3937531. (Accessed: 27 March 2023).

[13] N. Taconet, A. Méjean, C. Guivarch, 'Influence of Climate Change Impacts and Mitigation Costs on Inequality Between Countries', Climatic Change, vol. 160, pp. 15–34, 2020. [Online]. Available: https://doi.org/10.1007/s10584-019-02637-w.

[14] K. Dorkenoo, M. Scown, E. Boyd, 'A Critical Review of Disproportionality in Loss and Damage from Climate Change', WIREs Climate Change, vol. 13, p. e770, 2022. [Online]. Available: https://doi.org/10.1002/wcc.770.

[15] A. Jurgilevich, J. Käyhkö, A. Räsänen, S. Pörsti, H. Lagström, J. Käyhkö, et al., 'Factors Influencing Vulnerability to Climate Change-Related Health Impacts in Cities – A Conceptual Framework', Environment International, vol. 173, p. 107837, 2023. [Online]. Available: https://doi.org/10.1016/j.envint.2023.107837.

[16] B. B. Lin, A. Ossola, M. Alberti, E. Andersson, X. Bai, C. Dobbs, et al., 'Integrating Solutions to Adapt Cities for Climate Change', The Lancet Planetary Health, vol. 5, pp. e479–86, 2021. [Online]. Available: https://doi.org/10.1016/S2542-5196(21)00135-2.

[17] T. W. Rylenius, M. Hamza, 'The Fragmentation of Climate Change Adaptation – The Sweden Case', International Journal of Disaster Resilience in the Built Environment, 2024. [Online]. Available: https://doi.org/10.1108/IJDRBE-02-2023-0042.

[18] M. Sethi, L.-J. Liu, E. Ayaragarnchanakul, A. Suwa, R. Avtar, A. Surjan, et al., 'Integrated Climate Action Planning (ICLAP) in Asia-Pacific Cities: Analytical Modelling for Collaborative Decision Making', Atmosphere, vol. 13, p. 247, 2022. [Online]. Available: https://doi.org/10.3390/atmos13020247.

[19] J. Van Den Bergh, J. Castro, S. Drews, F. Exadaktylos, J. Foramitti, F. Klein, et al., 'Designing an Effective Climate-Policy Mix: Accounting for Instrument Synergy', Climate Policy, vol. 21, pp. 745–64, 2021. [Online]. Available: https://doi.org/10.1080/14693062.2021.1907276.

[20] L. K. Campbell, E. S. Svendsen, N. F. Sonti, M. L. Johnson, 'A Social Assessment of Urban Parkland: Analyzing Park Use and Meaning to Inform Management and Resilience Planning', Environmental Science & Policy, vol. 62, pp. 34–44, 2016. [Online]. Available: https://doi.org/10.1016/j.envsci.2016.01.014.

[21] Q. Dong, P. Zeng, X. Long, M. Peng, T. Tian, Y. Che, 'An Improved Accessibility Index to Effectively Assess Urban Park Allocation: Based on Working and Residential Situations', Cities, vol. 145, p. 104736, 2024. [Online]. Available: https://doi.org/10.1016/j.cities.2023.104736.

[22] T. M. Silva, S. Silva, A. Carvalho, 'Economic Valuation of Urban Parks with Historical Importance: The Case of Quinta do Castelo, Portugal',

Land Use Policy, vol. 115, p. 106042, 2022. [Online]. Available: https://doi.org/10.1016/j.landusepol.2022.106042.

[23] S. Suhardono, F. E. Safitri, F. Rachmalia, F. A. Rizky, D. H. Khairunnisa, I. Y. Septiariva, et al., 'Assessing the Ecological Function Effectiveness of Urban Parks in Surakarta City, Indonesia', Ecological Engineering & Environmental Technology, vol. 25, pp. 227–40, 2024. [Online]. Available: https://doi.org/10.12912/27197050/183790.

[24] M. Michel, J. Helnwein, 'Wie "fit" sind "klimafitte" Maßnahmen? – Methodenvorschlag zur interdisziplinären Bewertung von Wiener Parkanlagen', Energie und Klimawandel: Energie – Gebäude – Umwelt, vol. 26, Holzhausen, pp. 223–30, 2023.

[25] M. Pütz, S. Kruse, M. Butterling, 'Assessing the Climate Change Fitness of Spatial Planning: A Guidance for Planners', 2011.

[26] R. Seidl, 'Wie "klimafit" sind Österreichs Wälder, Flüsse und Alpenlandschaften?', 2012.

[27] Klima- und Energiefonds, 'KLAR! Programm. KLAR!-Serviceplattform Umweltbundesamt', 2023. [Online]. Available: https://klar-anpassungsregionen.at/klar-programm. (Accessed: 24 April 2023).

[28] M. Jurjonas, C. A. May, B. Cardinale, S. Kyriakakis, D. R. Pearsall, P. J. Doran, 'The Perceived Ecological and Human Well-Being Benefits of Ecosystem Restoration', People and Nature, 2023. [Online]. Available: https://doi.org/10.1002/pan3.10558.

[29] V. A. Masterson, S. Vetter, T. Chaigneau, T. M. Daw, O. Selomane, M. Hamann, et al., 'Revisiting the Relationships Between Human Well-Being and Ecosystems in Dynamic Social-Ecological Systems: Implications for Stewardship and Development', Glob. Sustain., vol. 2, 2019. [Online]. Available: https://doi.org/10.1017/sus.2019.5.

[30] K. Von Stackelberg, 'Ecosystem Resilience on Human Terms', Integr. Envir. Assess. & Manag., vol. 14, pp. 598–600, 2018. [Online]. Available: https://doi.org/10.1002/ieam.4073.

[31] B. J. Cardinale, J. E. Duffy, A. Gonzalez, D. U. Hooper, C. Perrings, P. Venail, et al., 'Biodiversity Loss and Its Impact on Humanity', Nature, vol. 486, pp. 59–67, 2012. [Online]. Available: https://doi.org/10.1038/nature11148.

[32] S. Díaz, J. Settele, E. S. Brondízio, H. T. Ngo, J. Agard, A. Arneth, et al., 'Pervasive Human-Driven Decline of Life on Earth Points to the Need for Transformative Change', Science, vol. 366, 2019. [Online]. Available: https://doi.org/10.1126/science.aax3100.

[33] E. C. Betley, A. Sigouin, P. Pascua, S. H. Cheng, K. I. MacDonald, F. Arengo, et al., 'Assessing Human Well-Being Constructs with Environmental and Equity Aspects: A Review of the Landscape', People and Nature, vol. 5, pp. 1756–73, 2023. [Online]. Available: https://doi.org/10.1002/pan3.10293.

[34] C. Wamsler, B. Wickenberg, H. Hanson, J. A. Olsson, S. Stålhammar, H. Björn, et al., 'Environmental and Climate Policy Integration: Targeted Strategies for Overcoming Barriers to Nature-Based Solutions and Climate Change Adaptation', Journal of Cleaner Production, vol. 247, 2020. [Online]. Available: https://doi.org/10.1016/j.jclepro.2019.119154.

[35] B. A. Bastien-Olvera, M. N. Conte, X. Dong, T. Briceno, D. Batker, J. Emmerling, et al., 'Unequal Climate Impacts on Global Values of Natural Capital', Nature, vol. 625, pp. 722–7, 2024. [Online]. Available: https://doi.org/10.1038/s41586-023-06769-z.

[36] A. G. Berberian, D. J. X. Gonzalez, L. J. Cushing, 'Racial Disparities in Climate Change-Related Health Effects in the United States', Current Environmental Health Reports, vol. 9, p. 451, 2022. [Online]. Available: https://doi.org/10.1007/s40572-022-00360-w.

[37] D. Chaudhry, 'Climate Change and Health of the Urban Poor: The Role of Environmental Justice', The Journal of Climate Change and Health, vol. 15, p. 100277, 2024. [Online]. Available: https://doi.org/10.1016/j.joclim.2023.100277.

[38] B. S. Ngcamu, 'Climate Change Effects on Vulnerable Populations in the Global South: A Systematic Review', Nat Hazards, vol. 118, pp. 977–91, 2023. [Online]. Available: https://doi.org/10.1007/s11069-023-06070-2.

[39] F. Green, N. Healy, 'How Inequality Fuels Climate Change: The Climate Case for a Green New Deal', One Earth, vol. 5, pp. 635–49, 2022. [Online]. Available: https://doi.org/10.1016/j.oneear.2022.05.005.

[40] C. A. Kukowski, E. E. Garnett, 'Tackling Inequality is Essential for Behaviour Change for Net Zero', Nat Clim Chang, vol. 14, pp. 2–4, 2024. [Online]. Available: https://doi.org/10.1038/s41558-023-01900-4.

[41] E. S. Parsons, A. Jowell, E. Veidis, M. Barry, S. T. Israni, 'Climate Change and Inequality', Pediatr Res, 2024. [Online]. Available: https://doi.org/10.1038/s41390-024-03153-z.

[42] M. Saint-Suppry Ceano-Vivas, M. J. Muñoz-Torres, J. M. Rivera Lirio, 'Revisiting the Relationship Between Sustainable Development and

Social Cohesion', SSRN Journal, 2014. [Online]. Available: https://doi.org/10.2139/ssrn.2521322.

[43] U. U. Baysal, P. C. Sutton, 'Desperately Seeking Sustainable Human Well-Being: A Review of Indicators, Concepts, and Methods', Ecological Civilization, vol. 1, p. 10004, 2024. [Online]. Available: https://doi.org/10.35534/ecolciviliz.2024.10004.

[44] M. Fortnam, P. Hailey, S. Witter, N. Balfour, 'Resilience in Interconnected Community and Formal Health (and Connected) Systems', SSM - Health Systems, vol. 3, p. 100027, 2024. [Online]. Available: https://doi.org/10.1016/j.ssmhs.2024.100027.

[45] D. Malerba, 'The Effects of Social Protection and Social Cohesion on the Acceptability of Climate Change Mitigation Policies: What Do We (Not) Know in the Context of Low- and Middle-Income Countries?', Eur J Dev Res, vol. 34, pp. 1358–82, 2022. [Online]. Available: https://doi.org/10.1057/s41287-022-00537-x.

[46] D. Schlosberg, 'Theorising Environmental Justice: The Expanding Sphere of a Discourse', Environmental Politics, vol. 22, pp. 37–55, 2013. [Online]. Available: https://doi.org/10.1080/09644016.2013.755387.

[47] Future Green City World Congress, 'Timetable Day 2', Future Green City World Congress, n.d. [Online]. Available: https://fgc2024.com/timetable-day-2/. (Accessed: 14 October 2024).

[48] F. Ferreira, L. Vasconcelos, J. C. Ferreira, 'Socio-Ecological and Economic Evaluation of Urban Parks – A Methodology Integrating and Articulating Diverse Components', Journal of Outdoor Recreation and Tourism, vol. 40, p. 100512, 2022. [Online]. Available: https://doi.org/10.1016/j.jort.2022.100512.

[49] S. Palabıyık, 'Evaluation of Small-Scale City Parks Based on Ecological Design Criteria', International Journal of Social Humanities Sciences Research (JSHSR), vol. 8, pp. 388–99, 2021. [Online]. Available: https://doi.org/10.26450/jshsr.2317.

[50] S. Zhang, H. Song, X. Li, S. Luo, 'Urban Parks Quality Assessment Using Multi-Dimension Indicators in Chengdu, China', Land, vol. 13, p. 86, 2024. [Online]. Available: https://doi.org/10.3390/land13010086.

[51] K. J. Bowen, S. Friel, 'Climate Change Adaptation: Where Does Global Health Fit in the Agenda?', Global Health, vol. 8, p. 10, 2012. [Online]. Available: https://doi.org/10.1186/1744-8603-8-10.

[52] Z. Guo, 'More Studies Needed on How Climate Change Affects Exercise Health Benefits', Nature, vol. 631, pp. 275–275, 2024. [Online]. Available: https://doi.org/10.1038/d41586-024-02249-0.

[53] G. Havenith, H. van Middendorp, 'The Relative Influence of Physical Fitness, Acclimatization State, Anthropometric Measures and Gender on Individual Reactions to Heat Stress', Eur J Appl Physiol, vol. 61, pp. 419–27, 1990. [Online]. Available: https://doi.org/10.1007/BF00236062.

[54] S. A. Morrison, 'Moving in a Hotter World: Maintaining Adequate Childhood Fitness as a Climate Change Countermeasure', Temperature, vol. 10, pp. 179–97, 2023. [Online]. Available: https://doi.org/10.1080/23328940.2022.2102375.

[55] M. Öberg, 'When Climate is Changing: Effects of Phenology and Local Climate on Individual Fitness', Doctoral Thesis, Department of Ecology, Swedish University of Agricultural Sciences, 2014.

[56] S. Schneider, A. Von Winning, F. Grüger, S. Anderer, R. Hoffner, L. Anderson, 'Physical Activity, Climate Change and Health—A Conceptual Model for Planning Public Health Action at the Organizational Level', Int J Environ Res Public Health, vol. 19, p. 4664, 2022. [Online]. Available: https://doi.org/10.3390/ijerph19084664.

[57] A. L. X. Cunha, R. M. D. Medeiros, N. L. D. L. Cavalcanti, R. M. D. Holanda, M. V. D. França, W. R. D. Araújo, et al., 'Potentiality of Sorghum's Climate Fitness for the State of Pernambuco, Brazil', RSD, vol. 9, p. e5589108706, 2020. [Online]. Available: https://doi.org/10.33448/rsd-v9i10.8706.

[58] F. S. Dobson, J. E. Lane, M. Low, J. O. Murie, 'Fitness Implications of Seasonal Climate Variation in Columbian Ground Squirrels', Ecology and Evolution, vol. 6, pp. 5614–22, 2016. [Online]. Available: https://doi.org/10.1002/ece3.2279.

[59] L. A. Ebensperger, D. S. Rivera, L. D. Hayes, 'Direct Fitness of Group-Living Mammals Varies with Breeding Strategy, Climate and Fitness Estimates', J Anim Ecol, vol. 81, pp. 1013–23, 2012. [Online]. Available: https://doi.org/10.1111/j.1365-2656.2012.01973.x.

[60] C. A. Fulton, K. E. Huff Hartz, N. W. Fuller, L. N. Kent, S. E. Anzalone, T. M. Miller, et al., 'Fitness Costs of Pesticide Resistance in Hyalella Azteca Under Future Climate Change Scenarios', Sci Total Environ, vol. 753, p. 141945, 2021. [Online]. Available: https://doi.org/10.1016/j.scitotenv.2020.141945.

[61] A. L. Hargreaves, S. F. Bailey, R. A. Laird, 'Fitness Declines Towards Range Limits and Local Adaptation to Climate Affect Dispersal Evolution During Climate-Induced Range Shifts', J Evol Biol, vol. 28, pp. 1489–1501, 2015. [Online]. Available: https://doi.org/10.1111/jeb.12669.

[62] J. G. Kingsolver, S. E. Diamond, L. B. Buckley, 'Heat Stress and the Fitness Consequences of Climate Change for Terrestrial Ectotherms', Funct Ecol, vol. 27, pp. 1415–23, 2013. [Online]. Available: https://doi.org/10.1111/1365-2435.12145.

[63] A. P. Møller, 'Interval Between Clutches, Fitness, and Climate Change', Behav Ecol, vol. 18, pp. 62–70, 2007. [Online]. Available: https://doi.org/10.1093/beheco/arl051.

[64] J. Cybulski, 'How to Build Your Climate Fitness', 2020.

[65] Rechnungshof Österreich, 'Wald im Klimawandel: Strategien und Maßnahmen', Wien, 2022.

[66] M. Pütz, S. Kruse, M. Butterling, 'CLISP Checkliste für Klimawandel-Fitness', 2011.

[67] S. Basar, R. Khalid, B. Rideout, L. Zoido, 'Planet Fitness', COMMON ACCOUNTS, 2019.

[68] Matadero Madrid, 'Climate Fitness', Matadero Madrid, n.d. [Online]. Available: https://www.mataderomadrid.org/en/schedule/climate-fitness. (Accessed: 16 October 2024).

[69] M. Borjabad, 'A Year-Long Program and Installation at Matadero Madrid - CLIMA FITNESS', COMMON ACCOUNTS, 2023. [Online]. Available: https://www.commonaccounts.online/. (Accessed: 16 October 2024).

[70] A. C. Lesnikowski, J. D. Ford, L. Berrang-Ford, M. Barrera, J. Heymann, 'How Are We Adapting to Climate Change? A Global Assessment', Mitig Adapt Strateg Glob Change, vol. 20, pp. 277–93, 2015. [Online]. Available: https://doi.org/10.1007/s11027-013-9491-x.

[71] Z. Yu, H. Gan, M. Xiao, B. Huang, D. Z. Zhu, Z. Zhang, et al., 'Performance of Permeable Pavement Systems on Stormwater Permeability and Pollutant Removal', Environ Sci Pollut Res, vol. 28, pp. 28571–84, 2021. [Online]. Available: https://doi.org/10.1007/s11356-021-12525-5.

[72] P. Malaviya, R. Sharma, P. K. Sharma, 'Rain Gardens as Stormwater Management Tool', in S. Shah, V. Venkatramanan, R. Prasad (eds.), Sustainable Green Technologies for Environmental Management,

Singapore: Springer, 2019, pp. 141–66. [Online]. Available: https://doi.org/10.1007/978-981-13-2772-8_7.

[73] E. Pochodyła, K. Glińska-Lewczuk, A. Jaszczak, 'Blue-Green Infrastructure as a New Trend and an Effective Tool for Water Management in Urban Areas', LO, vol. 92, pp. 1–20, 2021. [Online]. Available: https://doi.org/10.3097/LO.202192.

[74] A. Domlesky, 'Learning From Disaster: What Two Hurricanes Reveal About Ways to Design Public Space as Flood Infrastructure', J Clim Resilience Justice, vol. 1, pp. 66–77, 2023. [Online]. Available: https://doi.org/10.1162/crcj_a_00003.

[75] D. Mohaibesh, S. Monna, H. Qadi, R. Sokkar, 'Towards climate resilient residential buildings: learning from traditional typologies', Journal of Physics: Conference Series, vol. 2042, p. 012146, 2021. [Online]. Available: https://doi.org/10.1088/1742-6596/2042/1/012146.

[76] B. Li, W. Guo, X. Liu, Y. Zhang, P. J. Russell, M. A. Schnabel, 'Sustainable Passive Design for Building Performance of Healthy Built Environment in the Lingnan Area', Sustainability, vol. 13, p. 9115, 2021. [Online]. Available: https://doi.org/10.3390/su13169115.

[77] M. Gil-Baez, Á. Barrios-Padura, M. Molina-Huelva, R. Chacartegui, 'Natural ventilation systems to enhance sustainability in buildings: a review towards zero energy buildings in schools', E3S Web of Conferences, vol. 22, p. 00053, 2017. [Online]. Available: https://doi.org/10.1051/e3sconf/20172200053.

[78] A. Saifudeen, M. Mani, 'Adaptation of buildings to climate change: an overview', Frontiers in Built Environment, vol. 10, p. 1327747, 2024. [Online]. Available: https://doi.org/10.3389/fbuil.2024.1327747.

[79] V. Pellicer Sifres, 'Grassroots innovation for energy transition: Addressing citizen transformative practices for a new energy model', Universitat Politècnica de València, 2018. [Online]. Available: https://doi.org/10.4995/Thesis/10251/113073.

[80] D. Gielen, R. Gorini, N. Wagner, R. Leme, L. Gutierrez, G. Prakash, et al., 'Global Energy Transformation: A Roadmap to 2050', International Renewable Energy Agency, Abu Dhabi, 2019.

[81] X. Yao, K. Yu, X. Zeng, Y. Lin, B. Ye, X. Shen, et al., 'How can urban parks be planned to mitigate urban heat island effect in "Furnace cities"? An accumulation perspective', Journal of Cleaner Production, vol. 330, p. 129852, 2022. [Online]. Available: https://doi.org/10.1016/j.jclepro.2021.129852.

[82] T.-H. Su, C.-S. Lin, S.-Y. Lu, J.-C. Lin, H.-H. Wang, C.-P. Liu, 'Effect of air quality improvement by urban parks on mitigating PM2.5 and its associated heavy metals: A mobile-monitoring field study', Journal of Environmental Management, vol. 323, p. 116283, 2022. [Online]. Available: https://doi.org/10.1016/j.jenvman.2022.116283.
[83] S. Martín Muñoz, S. Elliott, J. Schoelynck, J. Staes, 'Urban Stormwater Management Using Nature-Based Solutions: A Review and Conceptual Model of Floodable Parks', Land, vol. 13, p. 1858, 2024. [Online]. Available: https://doi.org/10.3390/land13111858.
[84] H. P. Kodali, E. B. Ferris, K. Wyka, K. R. Evenson, J. M. Dorn, L. E. Thorpe, et al., 'The association of park use and park perception with quality of life using structural equation modeling', Frontiers in Public Health, vol. 11, 2023. [Online]. Available: https://doi.org/10.3389/fpubh.2023.1038288.
[85] D. Dushkova, M. Ignatieva, A. Konstantinova, F. Yang, 'Cultural Ecosystem Services of Urban Green Spaces. How and What People Value in Urban Nature?', Urban Ecology and Human Health, Springer, pp. 292–318, 2021. [Online]. Available: https://doi.org/10.1007/978-3-030-75285-9_28.
[86] D. Martens, Ö. Öztürk, L. Rindt, J. Twarok, U. Steinhardt, H. Molitor, 'Supporting biodiversity: Structures of participatory actions in urban green spaces', Frontiers in Sustainable Cities, vol. 4, 2022. [Online]. Available: https://doi.org/10.3389/frsc.2022.952790.
[87] R. Myalkovsky, D. Plahtiy, P. Bezvikonnyi, O. Horodyska, K. Nebaba, 'Urban parks as an important component of environmental infrastructure: Biodiversity conservation and recreational opportunities', Ukrainian Journal of Forestry and Wood Science, vol. 4, pp. 57–72, 2023. [Online]. Available: https://doi.org/10.31548/forest/4.2023.57.
[88] S. Hakimizad, S. R. Asl, M. M. Ghiai, 'A review on the design approaches using renewable energies in urban parks', International Journal of Renewable Energy Development, vol. 5, pp. 686–693, 2015.
[89] T. González, P. Berger, C. N. Sánchez, F. Mahichi, 'Sustainability of Urban Parks: Applicable Methodological Framework for a Simple Assessment', Sustainability, vol. 15, p. 15262, 2023. [Online]. Available: https://doi.org/10.3390/su152115262.
[90] W.R.W.A. Leduc, F.M.G. Van Kann, 'Spatial planning based on urban energy harvesting toward productive urban regions,' Journal of Cleaner Production, vol. 39, pp. 180–190, 2013. [Online]. Available: https://doi.org/10.1016/j.jclepro.2012.09.014.

[91] A. Tereci, Ç. Atmaca, 'Integrating Renewable Energy Systems into Urban Furniture for Recreational Spaces: A Design Proposal for Konya Adalet Park,' Gazi University Journal of Science, vol. 33, pp. 1–12, 2020. [Online]. Available: https://doi.org/10.35378/gujs.532828.

[92] J. Huebner, 'High Line in New York – An Unusual Park in the Centre of Manhattan,' IOP Conf Ser: Mater Sci Eng, vol. 471, p. 092085, 2019. [Online]. Available: https://doi.org/10.1088/1757-899X/471/9/092085.

[93] L. Bliss, 'The High Line's Biggest Issue—And How Its Creators Are Learning From Their Mistakes,' BloombergCom, 2017.

[94] H. Littke, R. Locke, T. Haas, 'Taking the High Line: elevated parks, transforming neighbourhoods, and the ever-changing relationship between the urban and nature,' Journal of Urbanism: International Research on Placemaking and Urban Sustainability, vol. 9, pp. 353–371, 2016. [Online]. Available: https://doi.org/10.1080/17549175.2015.1063532.

[95] R. Plunz, E. Moskalenko, High Line, Landscape Architecture Foundation, 2017. [Online]. Available: https://doi.org/10.31353/cs1250.

[96] Y. Dai, Y. Yin, Y. Lu, 'Strategies to Facilitate Photovoltaic Applications in Road Structures for Energy Harvesting,' Energies, vol. 14, p. 7097, 2021. [Online]. Available: https://doi.org/10.3390/en14217097.

[97] H. Hu, D. Vizzari, X. Zha, R. Roberts, 'Solar pavements: A critical review,' Renewable and Sustainable Energy Reviews, vol. 152, p. 111712, 2021. [Online]. Available: https://doi.org/10.1016/j.rser.2021.111712.

[98] S. Dubey, N.Y. Jadhav, B. Zakirova, 'Socio-Economic and Environmental Impacts of Silicon Based Photovoltaic (PV) Technologies,' Energy Procedia, vol. 33, pp. 322–334, 2013. [Online]. Available: https://doi.org/10.1016/j.egypro.2013.05.073.

[99] D. Hawila, A. Khalid, Renewable energy benefits: Leveraging local capacity for solar PV, Abu Dhabi: International Renewable Energy Agency, 2017.

[100] J. Li, J. Peng, C. Shuai, Z. Wang, F. Huang, M. Khayyam, 'Does the solar PV program enhance the social empowerment of China's rural poor?' Energy, vol. 253, p. 124084, 2022. [Online]. Available: https://doi.org/10.1016/j.energy.2022.124084.

[101] C. Behnke, T. Shelton, 'Powered by gentrification: The uneven development of residential rooftop solar in Atlanta, Georgia,' Energy

Research & Social Science, vol. 108, p. 103373, 2024. [Online]. Available: https://doi.org/10.1016/j.erss.2023.103373.
[102] G. Cranz, M. Boland, 'Defining the Sustainable Park: A Fifth Model for Urban Parks,' Landscape Journal, vol. 23, pp. 102–120, 2004. [Online]. Available: https://doi.org/10.3368/lj.23.2.102.
[103] S. Lang, J. Rothenberg, 'Neoliberal urbanism, public space, and the greening of the growth machine: New York City's High Line park,' Environ Plan A, vol. 49, pp. 1743–1761, 2017. [Online]. Available: https://doi.org/10.1177/0308518X16677969.
[104] M. Roach, 'The High Line Opened 15 Years Ago. What Lessons Has It Taught Us?' The New York Times, 2024.
[105] K.J. Black, M. Richards, 'Eco-gentrification and who benefits from urban green amenities: NYC's High Line,' Landscape and Urban Planning, vol. 204, p. 103900, 2020. [Online]. Available: https://doi.org/10.1016/j.landurbplan.2020.103900.
[106] N. Gulsrud, H. Steiner, 'When urban greening becomes an accumulation strategy: Exploring the ecological, social and economic calculus of the High Line,' Journal of Landscape Architecture, vol. 14, pp. 82–87, 2019. [Online]. Available: https://doi.org/10.1080/18626033.2019.1705591.
[107] A. Casellas, E. Dot-Jutgla, M. Pallares-Barbera, 'Artists, Cultural Gentrification and Public Policy,' Urbani Izziv, vol. 23, pp. s104–s114, 2012. [Online]. Available: https://doi.org/10.5379/urbani-izziv-en-2012-23-supplement-1-010.
[108] E. Jaffe, 'How Parks Gentrify Neighborhoods, And How To Stop It,' Fast Company & Inc., 2014.
[109] J.R. Wolch, J. Byrne, J.P. Newell, 'Urban green space, public health, and environmental justice: The challenge of making cities 'just green enough," Landscape and Urban Planning, vol. 125, pp. 234–244, 2014. [Online]. Available: https://doi.org/10.1016/j.landurbplan.2014.01.017.
[110] M. Hajdu, C. Simoneau, 'Chapter 15 - Communicating Science in a Policy Context to a Broader Audience', in Science for Policy Handbook, V. Šucha and M. Sienkiewicz, Eds., Elsevier, 2020, pp. 166–179. https://doi.org/10.1016/B978-0-12-822596-7.00015-2.
[111] C.R. Anderson, S.M. McLachlan, 'Transformative research as knowledge mobilization: Transmedia, bridges, and layers', Action Research, vol. 14, pp. 295–317, 2016. https://doi.org/10.1177/1476750315616684.

[112] R. Smiderle, S.J. Rigo, L.B. Marques, J.A. Peçanha de Miranda Coelho, P.A. Jaques, 'The impact of gamification on students' learning, engagement, and behavior based on their personality traits', Smart Learning Environments, vol. 7, no. 3, 2020. https://doi.org/10.1186/s40561-019-0098-x.

[113] M. Li, S. Ma, Y. Shi, 'Examining the effectiveness of gamification as a tool promoting teaching and learning in educational settings: a meta-analysis', Frontiers in Psychology, vol. 14, 2023. https://doi.org/10.3389/fpsyg.2023.1253549.

[114] M. Sailer, J.U. Hense, S.K. Mayr, H. Mandl, 'How gamification motivates: An experimental study of the effects of specific game design elements on psychological need satisfaction', Computers in Human Behavior, vol. 69, pp. 371–380, 2017. https://doi.org/10.1016/j.chb.2016.12.033.

[115] L-M. Putz, F. Hofbauer, H. Treiblmaier, 'Can gamification help to improve education? Findings from a longitudinal study', Computers in Human Behavior, vol. 110, p. 106392, 2020. https://doi.org/10.1016/j.chb.2020.106392.

[116] J. González-Martínez, M. Esteban-Guitart, C. Rostan-Sanchez, E. Serrat-Sellabona, M. Estebanell-Minguell, 'What's up with transmedia and education? A literature review', DER, pp. 207–222, 2019. https://doi.org/10.1344/der.2019.36.207-222.

[117] B. Tombleson, 'Transmedia learning: a literature review', Technology, Pedagogy and Education, vol. 33, pp. 255–269, 2024. https://doi.org/10.1080/1475939X.2024.2310681.

[118] A. Sánchez-Caballé, J. González-Martínez, 'Transmedia learning: fact or fiction? A systematic review (Aprendizaje transmedia: Â£realidad o ficción? Una revisión sistemática)', Culture and Education, vol. 35, pp. 1–32, 2023. https://doi.org/10.1080/11356405.2022.2121131.

[119] openPetition, 'Westbahnpark - Online-Petition', openPetition, n.d. https://www.openpetition.eu/at/petition/online/westbahnpark (accessed Nov. 4, 2024).

[120] C. Kirchmair, 'Frei:Raum Westbahnareal: Konversionsflächen als Ausgleich urbaner Verdichtung, dargestellt am Beispiel des 15. Wiener Gemeindebezirks Rudolfsheim-Fünfhaus', Masterarbeit, Universität für Bodenkultur Wien, 2019.

[121] N. Kaller, 'Westbahnpark: Stadt Wien macht keine klare Ansage', Wiener Zeitung, 2024.

[122] Institut für Stadt- und Regionalforschung der, Österreichischen Akademie der Wissenschaften, 'BLUEMAP: Map-based citizen knowledge on the use and development of Viennese blue spaces', 2024. https://www.oeaw.ac.at/isr/nachhaltige-stadtregion/bluemap-map-based-citizen-knowledge-on-the-use-and-development-of-viennese-blue-spaces (accessed Nov. 4, 2024).

[123] P. Kraus, 'Hilfe Hitze! Die Grünen Wien', 2022. https://wien.gruene.at/news/klimakrise/hilfe-hitze/ (accessed Nov. 4, 2024).

[124] Büro für lustige Angelegenheiten, 'Manifesto', n.d. http://www.bla.zone/manifesto/?lang=en (accessed June 29, 2023).

[125] WESTBAHNPARK.JETZT, 'Wir fordern den Westbahnpark!', n.d. https://www.westbahnpark.jetzt/(accessedJune29,2023).

[126] R. Rafiemanzelat, M.I. Emadi, A.J. Kamali, 'City sustainability: the influence of walkability on built environments', Transportation Research Procedia, vol. 24, pp. 97–104, 2017. https://doi.org/10.1016/j.trpro.2017.05.074.

[127] M. Lukacsy, C. Fendt, Statistisches Jahrbuch der Stadt Wien, 2022.

[128] City of Vienna, 'Urban Heat Vulnerability Index (UHVI) Wien - data.gv.at', 2023.

[129] M. Gandy, Concrete and Clay: Reworking Nature in New York City, First MIT paperback edition, Cambridge, Massachusetts, London: MIT Press, 2003.

[130] L. Lička, R. Tusch, U. Krippner, 'Park Politics', SPOOL, vol. 5, pp. 3–6, 2018. Available: https://doi.org/10.7480/spool.2018.2.3307.

[131] S. Pincetl, 'The Political Ecology of Green Spaces in the City and Linkages to the Countryside', Local Environment, vol. 12, pp. 87–92, 2007. Available: https://doi.org/10.1080/13549830601133102.

[132] A. B. Alvarez, 'New York City's High Line: Participatory Planning or Gentrification?', McNair Journal, pp. 1–14, 2012.

[133] K. Ascher, S. Uffer, 'The High Line Effect', Global Interchanges: Resurgence of the Skyscraper City, rlag Council on Tall Buildings and Urban Habitat, pp. 1–7, 2015.

[134] J. David, R. Hammond, 'High Line: the inside story of New York City's park in the sky', 1st ed., New York: Farrar, Straus and Giroux, 2011.

[135] L. L. Payne, A. J. Mowen, E. Orsega-Smith, 'An Examination of Park Preferences and Behaviors Among Urban Residents: The Role of Residential Location, Race, and Age', Leisure Sciences, vol. 24, pp. 181–98, 2002. Available: https://doi.org/10.1080/01490400252900149.

[136] A. J. Reichl, 'The High Line and the ideal of democratic public space', Urban Geography, vol. 37, pp. 904–25, 2016. Available: https://doi.org/10.1080/02723638.2016.1152843.
[137] P. L. Sacco, M. Tartari, G. Ferilli, G. Tavano Blessi, 'Gentrification as space domestication. The High Line Art case', Urban Geography, vol. 40, pp. 529–54, 2019. Available: https://doi.org/10.1080/02723638.2018.1502515.
[138] A. Rigolon, J. Németh, 'We're not in the business of housing: Environmental gentrification and the nonprofitization of green infrastructure projects', Cities, vol. 81, pp. 71–80, 2018. Available: https://doi.org/10.1016/j.cities.2018.03.016.
[139] J. Rothenberg, S. Lang, 'Repurposing the High Line: Aesthetic experience and contradiction in West Chelsea', City, Culture and Society, vol. 9, pp. 1–12, 2017. Available: https://doi.org/10.1016/j.ccs.2015.10.001.
[140] T. Slater, 'A Literal Necessity to be Re-Placed: A Rejoinder to the Gentrification Debate', Int. J. Urban Regional Res., vol. 32, pp. 212–23, 2008. Available: https://doi.org/10.1111/j.1468-2427.2008.00781.x.

Index

About the Editors

FH-Prof. Momir Tabakovic Ph.D., who holds a Doctoral of Science degree, is the head of two specialist areas at the University of Applied Sciences Technikum Wien: renewable energy technologies and climate-conscious buildings and neighbourhoods. He has extensive experience in research and teaching, having led numerous international projects in the fields of renewable energy, plus-energy neighbourhoods and integrated photovoltaics. His research is focused on the promotion of sustainable energy solutions and the improvement of cooperation between industry and universities. Currently, Tabakovic is spearheading an EU-funded initiative under the Horizon Europe programme, with a focus on enhancing the quality of life and reducing the cost of living in smart neighbourhoods.

Svitlana Alyokhina, Doctor of Technical Sciences, is a senior researcher and lecturer at the University of Applied Sciences Technikum Wien. With over 20 years of experience in scientific research, she specializes in industrial and applied thermal physics. Her expertise extends to numerical simulation, renewable energy technologies, and sustainable development. Throughout her career, she has contributed to numerous research projects, focusing on optimizing energy efficiency, integrating renewable energy sources, and advancing climate-friendly technological solutions. She has authored and co-authored multiple scientific publications and actively collaborates with academic and industrial partners to promote innovation in energy sustainability.